THE HAVEN-FINDING ART

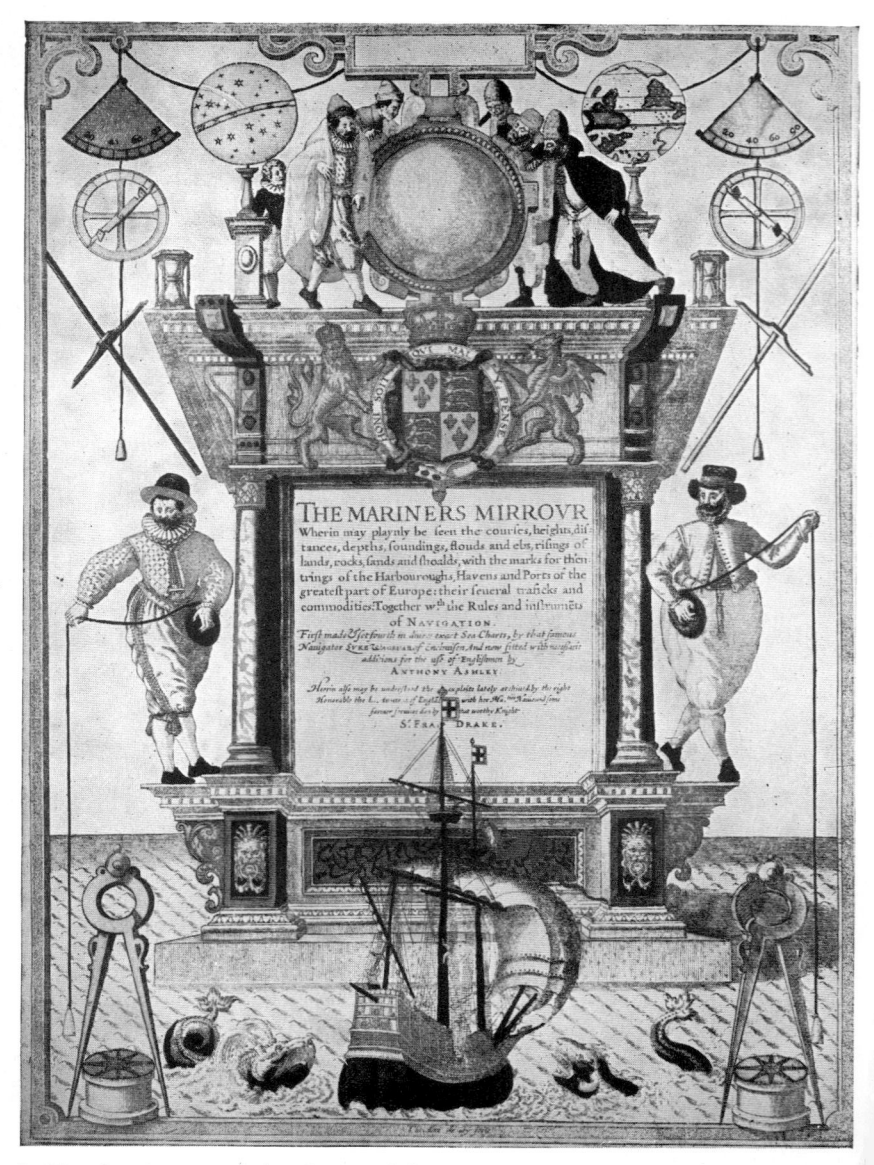

I. The frontispiece of the first English 'Waggoner' like its Dutch original emphasizes the important part played by soundings in the navigation methods of north-west Europe. The cross-staff, astrolabe and quadrant were of foreign introduction. The two bo'suns (right) carry their calls (whistles) on long chains, the two gentlemen officers (left) immediately under their chins.

THE
HAVEN-FINDING ART

A History of Navigation from Odysseus to Captain Cook

E. G. R. TAYLOR

With a Foreword by

COMMODORE K. St. B. COLLINS, R.N.

Hydrographer of the Navy, 1955-1960

Appendix by

JOSEPH NEEDHAM, F.R.S.

HOLLIS & CARTER

LONDON SYDNEY

TORONTO

FOR THE INSTITUTE OF NAVIGATION

Illi robur et aes triplex
   Circa pectus erat, qui fragilem truci
Commisit pelago ratem
   Primus

Horace, *Odes*, I. III. 9–12.

Oak and triple bronze must have girded the breast of him
who first committed his frail bark to the angry sea

Appendix to this edition © Cambridge University Press and
The Institute of Navigation 1971

Printed in Great Britain for
Hollis & Carter Ltd
9 Bow Street, London WC2E 7AL
and
The Institute of Navigation
1 Kensington Gore, London SW7
by The Stellar Press, Hatfield

First published 1956
Reprinted 1958
New augmented edition 1971

Contents

PART IV. INSTRUMENTS AND TABLES

PART V. TOWARDS MATHEMATICAL NAVIGATION

Plates

Plate I is reproduced by courtesy of the Royal Geographical Society; Plate II by courtesy of the University of Chicago Press; Plate III is reproduced from Greece in Photographs *by permission of Messrs Thames and Hudson; Plates IV, V and XXI are reproduced by courtesy of the Trustees of the British Museum; Plates VI and VIII are produced with permission from* Il Compasso da Navigare *edited by B. R. Motzo; and Plates IX, XII, XIII, XIV, XV, XVI, XVII, XVIII, XIX, XX, XXII, XXIII, XXIV and XXV are reproduced by courtesy of the Trustees of the National Maritime Museum. The block for fig. 11 (from a fifteenth-century Bible) was kindly lent by Messrs. Maggs.*

Figures

Foreword

COMMODORE K. St. B. COLLINS, R.N.

THERE are those who declare that when the Deity made man, on the sixth day, it was not intended that he should venture forth on the waters. Otherwise, they argue, he would have been fitted with waterwings, or flippers at least! If that is so, then the divine intentions have been very considerably modified since, without apparently any physical alteration to man, as the thousands and thousands who have ventured to sea testify.

There are those who have been launched upon it by force of circumstance, like Noah; or those who are born, live all their lives, and die upon it, like the little Chinese babies in the junks and sampans of Hong Kong harbour; there are those who have gone to sea to seek adventure and their fortunes, or simply to singe a beard.

Finally, and by far the greater number, there are those who have gone to sea, and will continue to do so as long as there is water to sail on, because it is in their blood and a love of it in their hearts. If you belong to this happy band, and nearly all Englishmen do, you will be enthralled with this book; and I venture to say that even if you are one of those unfortunates who must agree with the monk Alexander Neckam who (to quote the author) 'was strongly of the opinion that no one should go to sea except under extreme necessity' you must still be intrigued by this story of the evolution of the art of navigation from the days when 'a good sailor knew his ship and that was all', up through the years that saw the development of the compass, the chart, the Sailing Directions and nautical tables, the sextant, and last but by no means least, the chronometer.

Let it be said, despite all these things, that as much today as at any time in the past the 'good sailor' must know his ship and how she will react to wind and sea, if he is going to handle her competently and make the best use of the aids to navigating with which the mathematicians and the men of science have fitted her.

The life of a sailor has been changed over the years, to his greater comfort and his greater safety, as much by the designer and the ship-constructor as by the instrument-maker, the scientist and the mathematician. But the way of a sailor with his ship has not changed so much, and some of his methods not at all. So much in this book the modern navigator will recognize. He may still obtain the approximate altitude of a star by the span of his wrist (8°) or of his outstretched hand (18°) before looking for it in his sextant, just as his forbears did those hundreds of years ago. To this day the best method of finding a reported reef (I talk as a hydrographic surveyor—perish the thought that a prudent navigator would deliberately venture within miles of a reported danger!) is to make the latitude and then to search east and west through the longitude, because the latitude being easy to discover accurately is subject to less likelihood of error, by the reporting ship, than the longitude. This was exactly the method used by the old Portuguese and Spanish sailors who first journeyed to the Canaries and the Azores before those islands were accurately charted.

There is no navigator afloat today who will not approve the words, written some hundreds of years ago, 'navigating is not by chart and (magnetic) compass, but by the sounding lead!' Though the officer of the watch may be surrounded by all the accoutrements of a scientific age it still remains a basic fact, and one he will be wise not to forget for a moment, that if the draught of the ship exceeds the depth of water he is most assuredly aground!

In these pages you will find disproved once and for all that persistent myth that the first sailors navigated by 'hugging the shore'. Those words could never have been written by a sailor. Nothing is more fraught with peril, and therefore the more assiduously avoided on a little-known coast, than hugging the shore. The myth is based on the assumption that the mariner had neither the means nor the ability to find his way out of sight of land. Now this assumption is shown to be groundless. Also it takes no account of the extra sense and undoubted ability of the small-ship sailor to know where he is with remarkable accuracy without instruments or observations. It is a fact that many deep-sea fishermen have this ability in some measure today, and it may well have been more commonplace in days gone by. Nothing is more sure, by whatever means they achieved it, than that the sailors of all ages have navigated in deep waters.

This book brings the story up to the time of Cook. Looking back one finds with surprise how little the instruments changed from that date until the modern inventions of the gyro compass, the echo sounder, and radar, all within the last fifty years; in accuracy, a little; in workmanship, a little more perhaps; but in basic principles, not at all; and the finest exponent of their use was the greatest of all navigators and surveyors, Captain James Cook, R.N., F.R.S. 'Circumnavigator of the Globe, Explorer of the Pacific Ocean. He laid the foundations of the British Empire in Australia and New Zealand, charted the shores of Newfoundland and traversed the ocean gates of Canada both East and West.' So runs the inscription on his statue placed, appropriately enough, alongside the Hydrographic Department of the Admiralty by Admiralty Arch in the Mall.

K. St. B. Collins,
Commodore

Admiralty.
11 *November, 1955.*

Preface

The Haven-finding Art was the graceful phrase into which a notable Elizabethan translated the abrupt title of a Dutch work by Simon Stevin. The English version was made by Edward Wright, whose name every sailor knows, for it was he who in 1599 introduced the 'true chart' in the form still used in every ship today. Yet Wright was not a sailor; nor is the present writer. For to a seaman, making port is so familiar as hardly to be worth writing about, whereas the landsman has a passionate curiosity to learn just how a ship is steered with such confidence across the unpathed waters of the sea. It is out of such curiosity that this book has emerged, but a vigilant watch has been kept over it by a young navigator who has both a keen eye for lay blunders and a notable professional record and reputation of his own.

The author acknowledges with gratitude all the help received from Michael Richey, Executive Secretary of the Institute of Navigation, as inspirer and editor of the book, and offers her thanks besides to the following who have read the text or proofs and made useful comments: Mr. G. R. Crone, the Librarian and Map Curator of the Royal Geographical Society, Dr. G. E. R. Deacon, F.R.S., Director of the National Institute of Oceanography, Mr. René Hague, Mr. G. P. B. Naish of the National Maritime Museum, Mr. D. H. Sadler, Superintendent of H.M. Nautical Almanac Office; and to Miss Helen Wallis and Mr. R. A. Skelton of the British Museum Map Room. For being generous with answers to puzzling queries and for other favours she is also indebted to Mr. N. B. Marshall, of the Natural History Museum, Father Paul Grosjean, Bollandist, Commander Peter Scott, Mr. W. A. W. Scott of H.M. Nautical Almanac Office who calculated the star precession diagrams, and to many others. The quotation from Horace was the gift of Mr. Tracy Phillips.

In a story which begins with Odysseus and ends with Captain Cook there can hardly fail to be misinterpretations and errors, but for these the author alone is responsible.

E. G. R. T.

PART ONE

INTRODUCTION

I

Signs in the Sky

How did the ancient navigator, with neither compass nor chart, set and keep course for his port of destination? There is no direct answer to this question, for the sailor was a craftsman, learning as a youth how to pilot his ship by working beside his master. Nothing was written down. The techniques of the arts and crafts were, indeed, of no concern to the clerks and scribes who alone had the art of writing. Yet it is from their hieroglyphic record of events that we accidentally know how ancient sailing is, for their symbol for a foreign ship showed it with a square sail. Such ships were visiting the Red Sea ports of Egypt at least two thousand years before our era. Moreover, fine seal-engravings, showing large ships driven by sails and oars, survive from ancient Crete, the great Mediterranean sea-power of the second millennium B.C. It is not, however, until we get a written narrative literature that we can glean anything of the conduct of such a ship. And even then we must rely upon stray phrases, dropped now and then in the telling of a sea-story, or upon the report of some passenger aboard, who relates what he saw, or thought he saw, the sailors doing. In the Acts of the Apostles, for example, St. Paul's shipwreck is vividly described by an eyewitness, who tells us that when for many days neither Sun nor stars had been visible from the driving ship, the crew abandoned all hope. For the Sun by day, and the stars by night, served the helmsman as compass for his bold sailing during the three thousand years or more which elapsed before he knew the magnetic needle. And for chart he relied upon his visual memory and experience of the coastal sky-lines. Even today, of course, since the ultimate sources of time-keeping and position-finding are the heavenly bodies, the sailor must look up at the sky. But so long and so far has a chain of experts—professional astronomers, mathematicians, almanac-makers, instrument-makers and so forth—separated the ordinary man from first-hand observation that he has ceased to think beyond the actual clock, time-signal, map, calendar, or whatever it may be that 'tells' him what he wishes to know.

The traveller by land, when at a loss for his way, has only to ask some passer-by which road or track to take. And because he himself never has occasion to consult the sky, he finds it impossible to imagine how the seaman makes his confident way across the pathless and unpeopled ocean waters. Historians, confronted with the fact that voyages were actually made, coined the phrase that, before he had the magnetic compass, the sailor 'hugged the shore', and crept coastwise from port to port by roundabout routes. It is true, of course, that there was coastwise sailing, but every sailor has a wholesome dread of being driven on to a lee-shore, and stands well out to sea to avoid the dangers of hidden rock and sand-bank, of breakers and tide-rips, which are characteristic of inshore waters. In some current Admiralty sailing directions, for example, the sailor making for the south-west past Brittany is told concisely: 'Ushant must not be sighted', for if he is as near in as that, the powerful on-shore set may sweep him on to the rocks.

Evidence for early open-sea and cross-sea voyages is plain enough, even before the days of sail. Finds of ancient Irish gold ornaments of the bronze age show that such articles were being traded by sea to France, and were also carried across Scotland and then shipped over the North Sea to Denmark. Even more convincing are the massive stone foundations, which can still be seen, of the seaport built on the south shore of Crete by the pre-Greek Minoans (Fig. 1). It stood at the end of a road coming directly from the royal city of Cnossus on the north coast, and was clearly designed for direct trade with Africa and Egypt. Even the shortest course from shore to shore would take two days with a fair wind, and it is therefore clear that the Cretan pilot could keep course by night as well as by day quite out of sight of any land, steering only by stars and Sun. The Sun, appearing over the horizon at one point, and disappearing at another, has given to all mankind the first general ideas of direction, and consequently the first names for directions, other than purely local ones. East and west, orient and occident, lay at the opposite ends of the Earth where bright day was divided from dark night. In Homer's

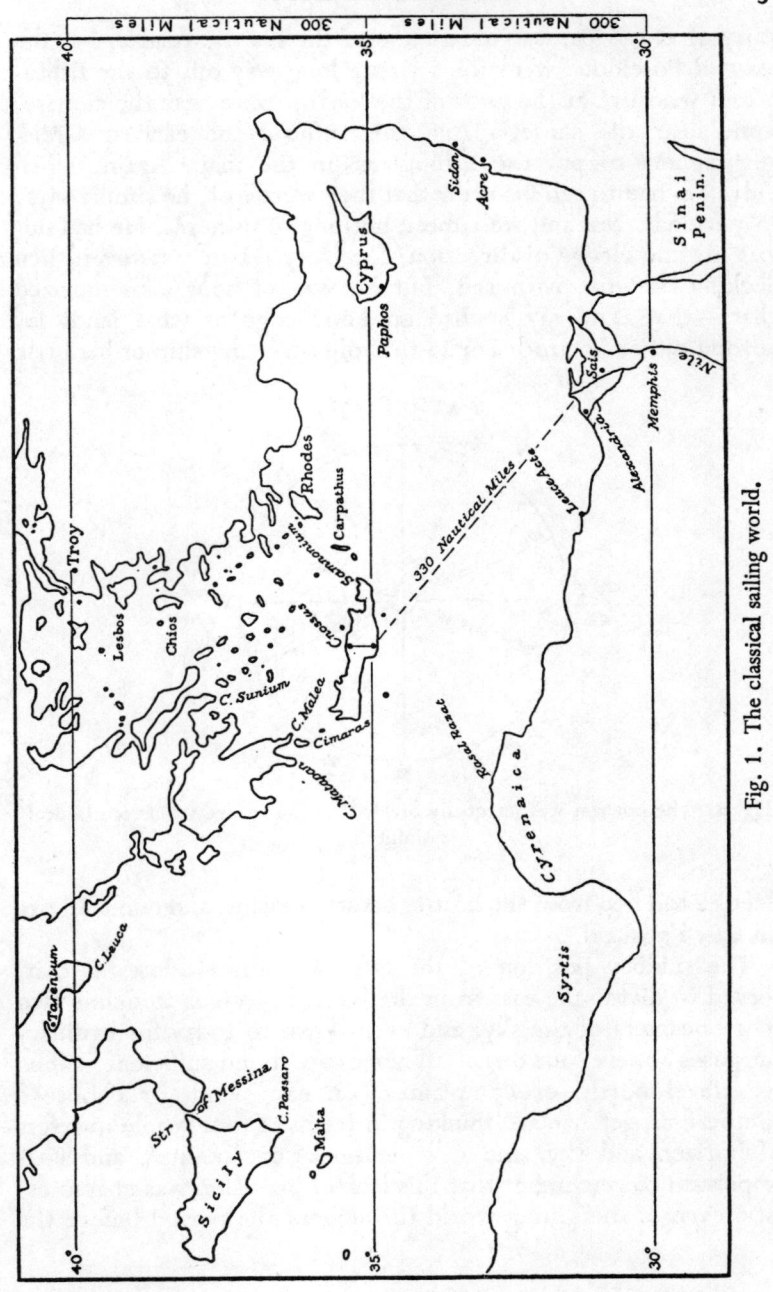

Fig. 1. The classical sailing world.

story of Odysseus, our oldest tale of the sea, he relates that the sea-god Poseidon 'went for a visit a long way off, to the Ethiopians: who live at the ends of the Earth, some near the sunrise, some near the sunset'. And this sufficed the earliest Greek geographers to put the Ethiopians on the map. Again, when Odysseus had to tell his crew that they were lost, he simply says: 'My friends, east and west mean nothing to us here.' He had not lost his knowledge of direction, for 'rosy Dawn' was even then flecking the east with red, but he was without a memorized chart—that is to say he had no knowledge of what lands lay beyond either horizon. For in the sole surviving ship of his little

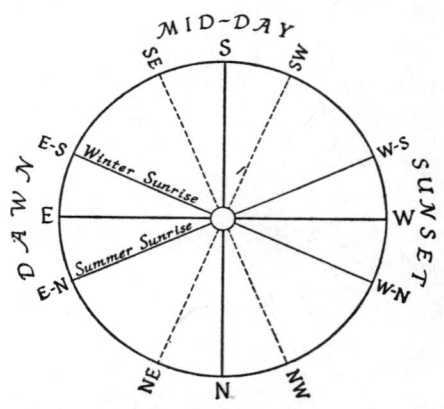

Fig. 2. The horizon was unequally divided by the twelve winds of classical antiquity.

fleet he had fled from the hostile Laestrygonians, and come ashore on Circe's island.

The midday position of the Sun, and the shadows it cast, served to divide the east from the west, as well as to define two more quarters of the sky, and even down to today for ordinary purposes a mere four direction names are found sufficient. When we travel north, or complain of an east wind, or choose a southern aspect, we are thinking in terms of four whole quarters of horizon and sky, and not of 'lines' or 'points', and it is important to remember that this lack of precision was characteristic even of the sailing world throughout the period before the magnetic needle. A man might speak of 'between west and north', or 'a little north of east', but he was not thinking in terms of the bisection of angles, or of a precise arc of the horizon, since he had no geometrical concepts. The first refinement in dividing the horizon of which we know (leaving aside of course the ideas of the tiny circle of the astronomers) was that made among the Greeks, who named the directions of summer and winter sunrise or sunset, as distinct from the east and west directions in the spring and autumn seasons of equal days and nights (Fig. 2). Since the Mediterranean world lay roughly between Lat. 30° and 40°, these directions expressed in mathematical terms would be from 27° to 31° or thereabouts to the

Fig. 3. The eight separately named rhumbs of Mediterranean sailors.

north and south of the east–west line. They marked off, that is to say, about one-third of each quadrant, and were looked upon as 'companions' of the equinoctial east. It was natural, therefore, to give similar companions to the two directions north and south, so that from about the fourth century B.C. there was a system of twelve directions, used side by side with an eightfold system which arose from roughly halving the four quadrants (Fig. 3). The twelvefold system, however, was closely bound up with wind-directions rather than Sun-directions, and will therefore be described in the next section. There was, besides, another and quite different division of the four quadrants which came into use at a much later date, among the Vikings. These, as a northern people, were accustomed to sunrise and sunset positions which altered very rapidly, and embraced almost the whole sweep of the horizon. East and west were therefore much less easy to pick out and name than north and south, and the latter, besides, were in high latitudes much more vital in daily life. The shores of

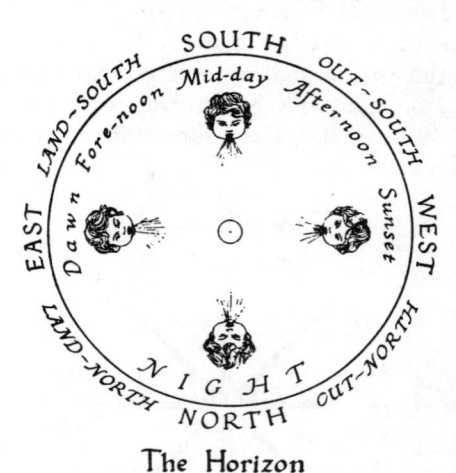

The Horizon

Fig. 4. The quarters of the horizon named by the Norsemen.

Denmark and Scandinavia, as it happens, run roughly from south to north, and it was very natural to distinguish directions east of the meridian as 'land-north', and 'land-south', while the corresponding seaward directions were termed 'out-north' and 'out-south', meaning roughly what we should call north-west and south-west (Fig. 4).

The night-sky, to those whose avocations demanded that they should watch it—sailors, shepherds and astronomers—gave quite as clear indications of direction as the day. The appearances and disappearances of the stars divide the whole horizon into an eastern and a western half, and some astronomers considered the north–south meridian line merely as the boundary between east and west. It was the line along which each star, in its turn, and at its own moment, ceased to ascend and began to descend, on which, that is to say, it culminated. The whole northern quarter, however, is picked out unmistakably for everyone by a brilliant group of the seven stars—the Bear, Plough or Wain as they are variously called. These neither rise nor set, but circle round and round, like a wheel about its axle, and so serve to define the stillpoint, the pommel or pole of the sky. For the Greeks the word for Bear—Arctos—was also the word for north, while the Latins called the north Septentrio, said to derive from *septem triones* the seven plough-oxen. This star group played a prominent part in Egyptian astronomy, which goes back perhaps as far as 5000 B.C., while Homer reciting his poetry some eight or nine hundred years B.C. expected all his hearers to know it, as did the Hebrew prophet Amos whose book belongs to the eighth century B.C. Its circumpolar character, too, was remarked upon by Homer, who says that the constellation 'wheels round and round where it is, watching Orion, and alone of them all never takes a bath in the Ocean'. His phrase 'alone of them all' may be noted, for it is a reminder that in his day the Lesser Bear, because it is so much harder to pick out, had not been distinguished as a star group by the Greeks. Cassiopeia, too, a prominent circumpolar constellation of today, was then much farther from the celestial pole, and disappeared nightly 'beneath the ocean', that is to say beneath the waters which the early Greeks deemed to encircle the whole horizon of the Earth (Fig. 5).

Orion the Hunter, whom Homer describes as being watched cautiously by the Bear, moves across the southern quarter of the sky, and when he rises his three-starred belt catches the eye at once. But as only half his course or less is above the horizon, he will only be seen during the season when his rising does not take place in daylight. Like Homer, the prophet Amos coupled him with the Bear, in the opposite part of the sky, for Amos is poetically naming the four quarters of the firmament when he writes: 'Seek him that maketh the Seven Stars and Orion, and turneth the shadow of death into the morning, and maketh the day dark with night.' The point in the heavens about which the whole starry pattern appears to wheel is not in actual fact a fixed point; it shifts through the centuries extremely slowly along an elliptical path which astronomers can trace out, so that we know how the stars appeared to observers in relation to the pole during different periods of history. The Great Bear,

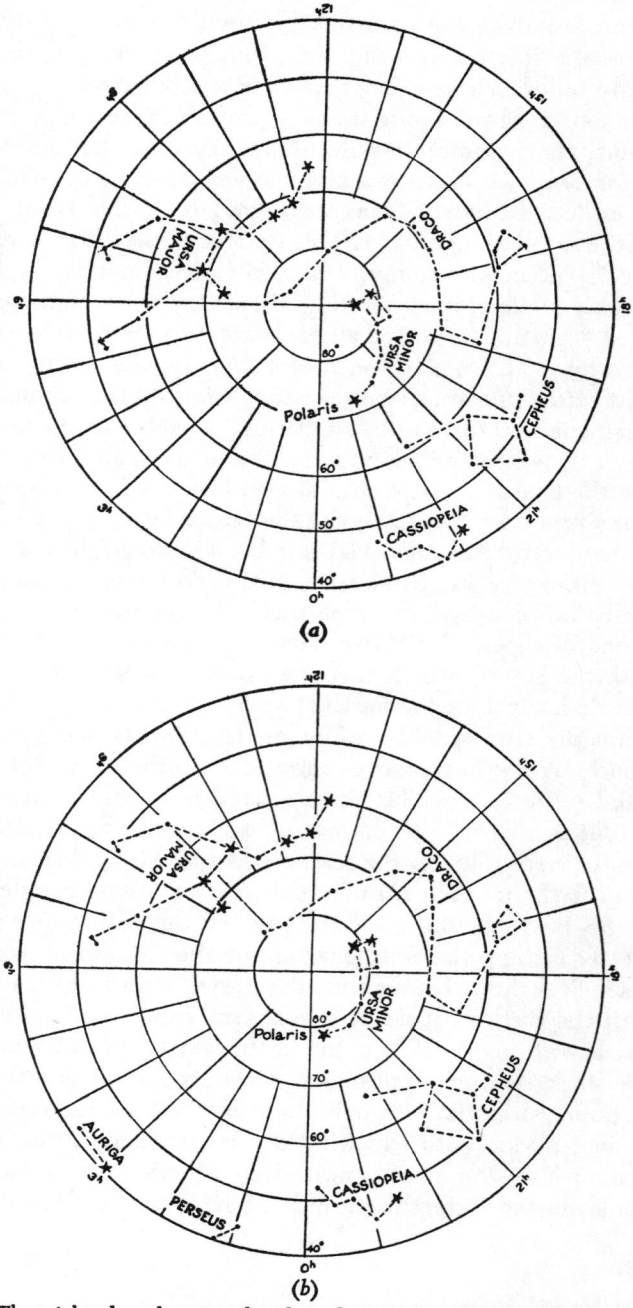

Fig. 5. The night sky, showing the altered positions of the circumpolar stars between 1000 B.C. and the present day.

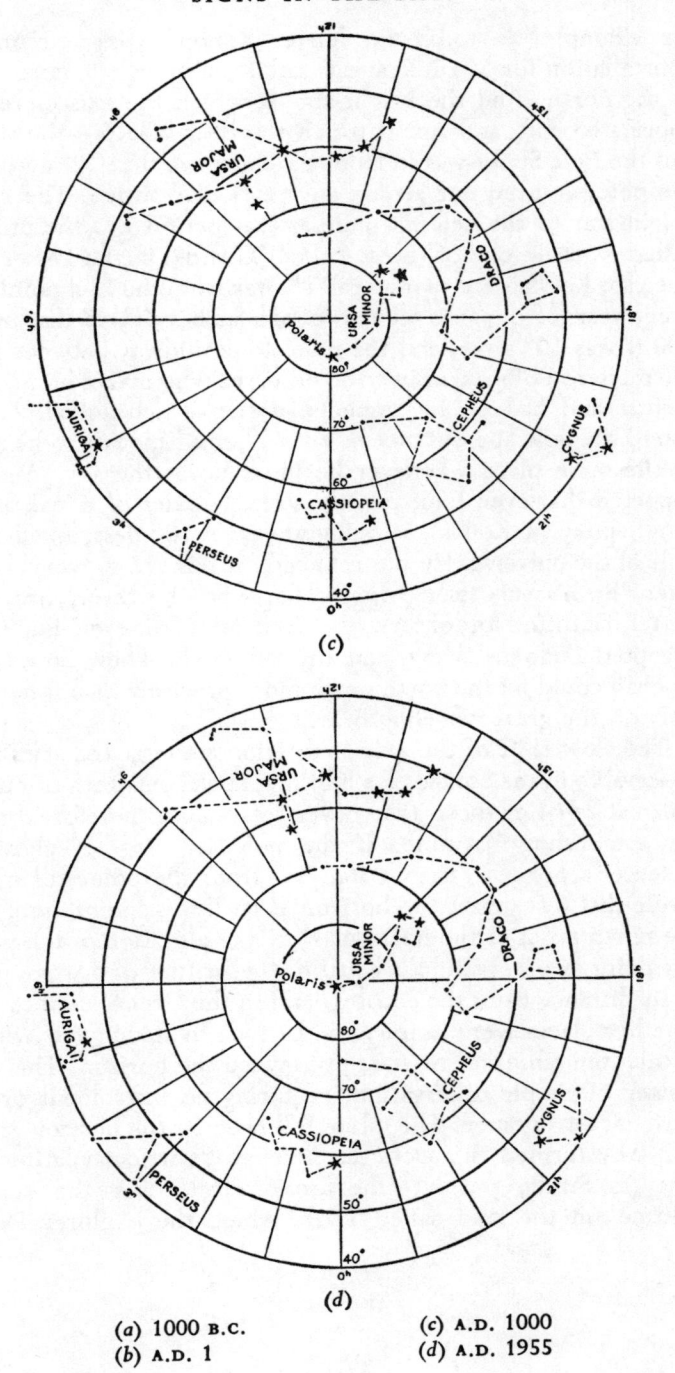

(c)

(d)

(a) 1000 B.C.   (c) A.D. 1000
(b) A.D. 1   (d) A.D. 1955

for example, is today no longer a non-setting circumpolar constellation for Mediterranean sailors, although it is so for us in the north. And the bright star at which the axis of rotation appears to end, and round which the Bear wheels—the star we call the Pole Star—was in Homer's day more than 12° away from the pole, and did not attract any particular notice. The nearest bright star to the celestial pole at that period was the one now called Kochab, one of the so-called 'Guards' in the Lesser Bear. Even so, Kochab was as much as 7° away from the 'nul point', and every year left it some seconds of arc farther off, so that by A.D. 400 it was 10° away, and the pole stood midway between it and the modern Pole Star. But to an observer who makes no measurements, and had no instruments, a star which actually circles round the pole at a distance of even 7° or 8° appears to be always in the same place whenever he looks up at the sky. And it is reported that even Eudoxus, the great founder of Greek astronomy, spoke of Kochab as being always motionless, fixed at the pole of the universe. He was rebuked for this error, two centuries later, by his successor Hipparchus, who, however, had much better facilities for observation. But after all even Euclid had accepted Eudoxus' view, and any sailor who knew how to find Kochab could fix the north much more precisely than if he relied only on the great wheeling Bear.

The slow shift of the axis of rotation amongst the stars is not observable in the course of a lifetime, and the pattern of circumpolar stars—i.e. those that never set—could therefore be used for establishing latitude. If the pole is, say, 35° above the observer's horizon, then a star 35° from the pole will make a circle that just grazes the horizon at its lowest point, actually at the north point of the horizon. Now the elevation of the pole at any point of observation is equal to the latitude of that point, i.e. to its distance from the equator, and so the Greeks used to judge whether places were in the same latitude by finding out whether at each the same star or stars just grazed the horizon. This was a matter of simple observation, requiring no instrument or professional astronomer. The circle followed by the horizon-grazing star they termed the arctic circle of that particular latitude, so that (as Strabo put it) 'the arctic circle was the summer solstice' in the land called Thule which the explorer Pytheas described. By this he meant that all the stars as far as the celestial tropic of Cancer were circumpolar stars, which makes the latitude of Thule that of the arctic circle in its later sense of a fixed circle, the one where the midnight Sun first appears. Whether sailors judged their position by the horizon-grazing stars cannot be known, but astronomers used them to measure the Earth. According to Eudoxus and Posidonius the bright star Canopus in the southern celestial hemisphere could only just be seen on the southern horizon at Rhodes, whereas it rose and culminated at an altitude of $7\frac{1}{2}°$ in Alexandria. Now according to sailors the distance between these two ports was 4000 stadia from north to south (others said it was 5000), and hence by simple proportion the number of stadia corresponding to one degree of difference in the star altitude—i.e. to one degree in latitude, could be found. The figures were faulty, and the results therefore only approximate—either 500 or 700 stadia to a degree in round numbers—but the principle was sound.

For the stationary observer the stars appear at precisely the same point on the horizon every day, and culminate as they cross the meridian at precisely the same altitude. But they rise, culminate and set almost four minutes earlier every twenty-four hours according to our time reckoning, which is by the Sun's return to the meridian. In actual fact it is the Earth's motion along its orbit that makes Sun time four minutes late on star time every day, and in the course of the year twenty-four hours are lost and the times come together again. Hence the times of star-rise can be used to mark the seasons, and in ancient times men watched for the first appearance of some well-known star at dawn, that is to say in morning twilight which (as Chaucer reminds us) was included like the evening twilight in the common working day. The rising of a star with the Sun was termed its heliacal rising, and while at first it would be almost immediately obscured by the coming daylight, each day it would rise earlier and be visible for a longer and longer period, until at length it shone all night, rising at sunset and setting at sunrise. Thus the heliacal rising of the Pleiades in May proclaimed that the season for harvesting the grain (grown in winter in the Mediterranean) had arrived, and when in November this constellation was setting at sunrise, everyone knew that the time to plough and sow had arrived. It was the rising of Sirius with the Sun that announced to Egypt the time to expect the rising of the Nile. Sirius is the brightest of all the stars and is the Great Dog who follows Orion. This hunter's splendour derives from his association with four out of the nine most brilliant stars in the sky. Besides Sirius he has the Lesser Dog, Procyon, behind him, while Betelgeuse shines in his shoulder and Rigel in his foot. As well known to the ancients as Orion was the solitary Arcturus, the driver of the Plough, or Wain, who is easily picked out by following the sweep of the Great Bear's tail. Ascending as he did nearly to the zenith in Mediterranean skies, he was as familiar to the Hebrews as he was to Homer and the Greeks. When Job, for example, describes the Creator 'which alone spreadeth out the heavens', he adds 'which maketh Arcturus, Orion and Pleiades'. And when 'the Lord answers Job out of the whirlwind', he says: 'Canst thou bind the sweet influences of Pleiades, or loose the bands of Orion? Canst thou bring forth Mazzaroth [the circling Zodiac] in his season? or canst thou guide Arcturus with his sons?' Two other bright and lovely stars were named in olden days which only the astronomers knew to be actually appearances of the white planet Venus. These were the Morning Star and the Evening Star, Lucifer and Hesperus, who marked the coming and the closing of the day, the east and the west. For Venus is never far from the Sun. The Moon, too, has always been a sky-sign, passing from east to west, although too variable to be of much help to sailors. Yet it was the sailor who early discerned the relation of her waxing and waning to the rise and fall of the tide, and it was to pin down the motions of the Moon that the Royal Observatory at Greenwich was founded.

The twelvefold direction system which has been mentioned was bound up with the pattern and character of the winds. For the wind brings the weather. And the kind of weather it brings depends very closely upon the quarter from which it blows. It follows that the 'feel' of a wind gives a rough indication of direction, and it is not surprising that the names given to well-recognized winds were also used simply as direction names, and that a division of the compass of the horizon in terms of winds supplemented or even superseded that in terms of Sun and stars. The simplest observation is that cold air comes from the north, warm air from the south, so that Boreas, the uncomfortable northerly wind, was also a name for north, and Notus, the south wind, meant also the south. Easily distinguishable, too, in the Mediterranean world, was the mild breeze, bringing showers of rain, which blew from the direction of the sunset and was called Zephyr. On the other hand the wind from the east, Apeliotes, felt dry and cool. For those who thought in terms of four directions only, this was enough. But for those—including philosophers and sailors—who observed more closely, there was a clear distinction between a moist north wind bringing snow and its companions on either quarter. Dry, piercingly cold blasts came from the arc of the horizon to the east of north, and this was the true Boreas. On the other hand if the north wind had a westerly element it was strong and blustering, bringing rain, hail and thunder. It was then called Argestes. The south wind, too, was observed to be dry and hot when it came from the direction towards the winter sunrise, moist and hot when it came from towards the winter sunset. Thus an eightfold wind-direction system was recognizable, and it is this that was displayed on the Tower of Winds still to be seen at Athens (Plate III). Strabo tells us that some people considered that Boreas and Notus were the only principal winds, all others being modifications of them, but he goes on to say that this is not held correct by the best authorities, such as Aristotle Timosthenes, King Ptolemy's Chief Pilot, and Bion the astrologer who, according to Posidonius, named eight. But in fact there were those who gave the north and south winds each two companions and joined these six to the six directions derived from the seasonal sunrise and sunset, and so produced a twelvefold system, which became the standard system in classical literature. However, there is good reason to suppose that mariners used only the eightfold set of wind-names, or wind-rose, as it was later termed. Nevertheless there was considerable confusion arising from names implying particular directions, since some of these had been derived from the names of the countries in which the particular winds were supposed to take their rise. The Greeks spoke, for example, of a Thracian wind, presumably a north-easter, and Herodotus tells us of an easterly wind which was called the Hellespontian. The Athenians called the north-west wind Sciron, after the hills from which it came. Such names lost their original topographical significance when they came into more general use, just as the Viking land-north and out-north did after the colonization of Iceland and Greenland, and the exploration of the coast of America. Meanwhile we find that during the Viking age the Anglo-Saxons had already devised or adopted the system, later to be accepted by sailors the world over, of making a simple combination of the four basic direction names, north, south, east and west, in order to express intermediate directions and give winds their names at the same time. King Alfred used the terms north-east ('northan-ostan'), south-east, north-west, south-west, and this eightfold system (adding the cardinals) appears to have had no further extension until the appearance of the magnetic needle.

In spite of the saying that the wind 'bloweth where it listeth', there is in actual fact a recognizable pattern of winds over the globe, although it is a pattern of prevailing or most frequent winds only. These may drop, or be reversed in a fashion that appears capricious: only, however, because of the insufficiency of our knowledge. Most of our observations are of the lower few hundred feet of the atmosphere, where the wind is locally modified by the contours and nature of the ground. In the Strait of Gibraltar for example, with its high-standing shores, all winds are turned to run either east or west, while round any lofty promontory there are developed local gusts and eddies with directions all their own which may be violent and dangerous to small ships.

Broadly speaking, the steadiest winds (which encourage sailing) are to be found in low latitudes, between about Lat. 30° and the equator in either hemisphere, while the most variable winds are to be found in high latitudes (Fig. 6). Gigantic masses of air are continually settling down over the regions lying in or about Lat. 30°, so that there the barometer stands high, and as there is little horizontal air movement, the observer finds himself in a belt of calms—the horse latitudes of sailors. From both margins of the horse latitudes air flows out strongly, producing the steady trade winds on the equatorward side, and the stormy westerlies on the poleward side. The two sets of trade winds—from north-east and south-east respectively—converge and meet in the neighbourhood of the equator, where they halt one another. Some of the air flows away eastwards, but much of it piles up in rain-giving convection currents and flows back at high level towards the horse latitudes. In the narrow belt of ascending air between the trades there are only light and variable winds or long calm spells near the surface. This is the 'doldrums' where sailors were becalmed for weeks. The air which flows or blows out on the poleward

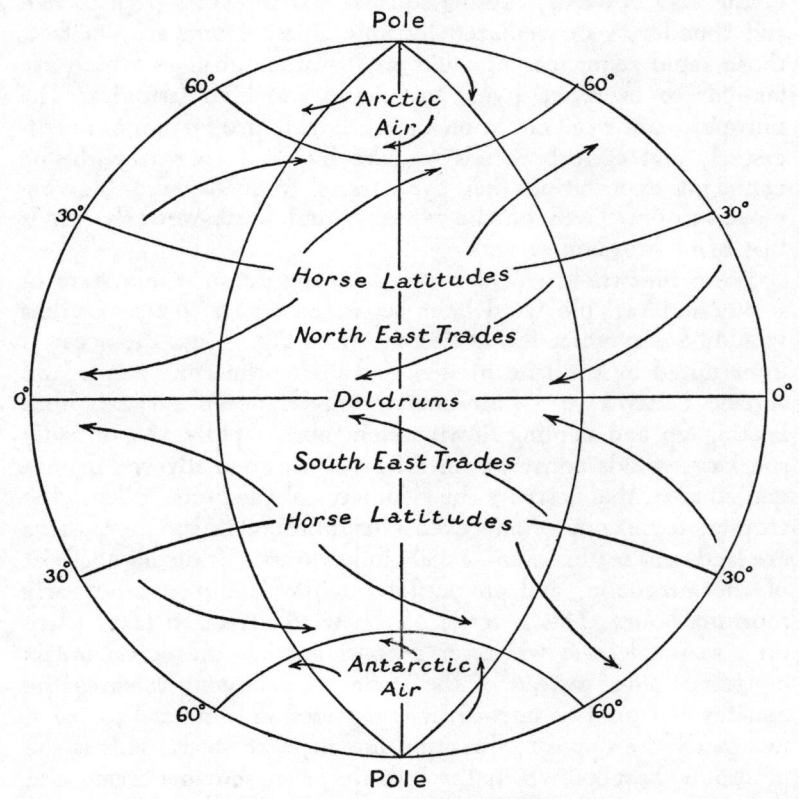

Fig. 6. The scheme of the wind belts.

side of the horse latitudes—as south-westerly to westerly winds in the northern hemisphere, north-westerly to westerly winds in the southern hemisphere—sooner or later encounters heavy masses of cold, dry air which are spreading out from the polar regions. This meeting of very dissimilar bodies of air is a very different matter from the convergence of the trade winds, which roughly resemble one another in their temperature, moisture and rate of flow. The warmer, lighter air from lower latitudes, impinging strongly on a colder, more inert mass, will flow upwards over it, producing heavy cloud and steady rain, while a tongue of polar air impinging on the warm mass will thrust it out of the way upwards, causing squalls, and showers, perhaps hail and thunder, soon replaced by blue skies. There are, in fact, those rapid sequences of wind and weather changes which are familiar to everyone living in middle and high latitudes. The European sailor can count on a considerable proportion of north-easterly and easterly winds to take him out to sea, with the confident expectation that even more frequent, and blowing more strongly, will be the westerly and south-westerly winds that bring him home.

Were the whole world covered by the ocean, the system of steady and variable wind belts separated by the belts of calms would be as symmetrical as that shown in Fig. 6. But the ocean is interrupted by the land masses, and under the same Sun a land surface behaves very differently from an ocean surface, both heating up and cooling down much more rapidly. As, broadly speaking, winds converge on a heated area and diverge from a cooled area, this destroys the symmetry of the belts. Within the tropics, for example, and even outside them in summer, there are land- and sea-breezes—a daily inflow of sea air during the heat of the afternoon, and an outflow in the late night and early morning hours. This reversal of the wind direction takes place on a seasonal scale wherever a continent has the ocean on its equatorial side. Instead of the trade wind flowing towards the equator as would be normal, it is replaced in summer by a wind in exactly the opposite direction flowing landwards. This is the monsoon, best known in the Indian Ocean, but occurring also in south-east Asia, north Australia and the Guinea Gulf as well. The to-and-fro movement of the air, north-east alternating with south-west, which can be relied upon between India and East Africa has undoubtedly played a great part in fostering overseas navigation there since very early times (Fig. 7a).

Another fateful consequence of the rapid way in which a land surface changes temperature is the development of masses of intensely cold air over the northern continents in winter, and

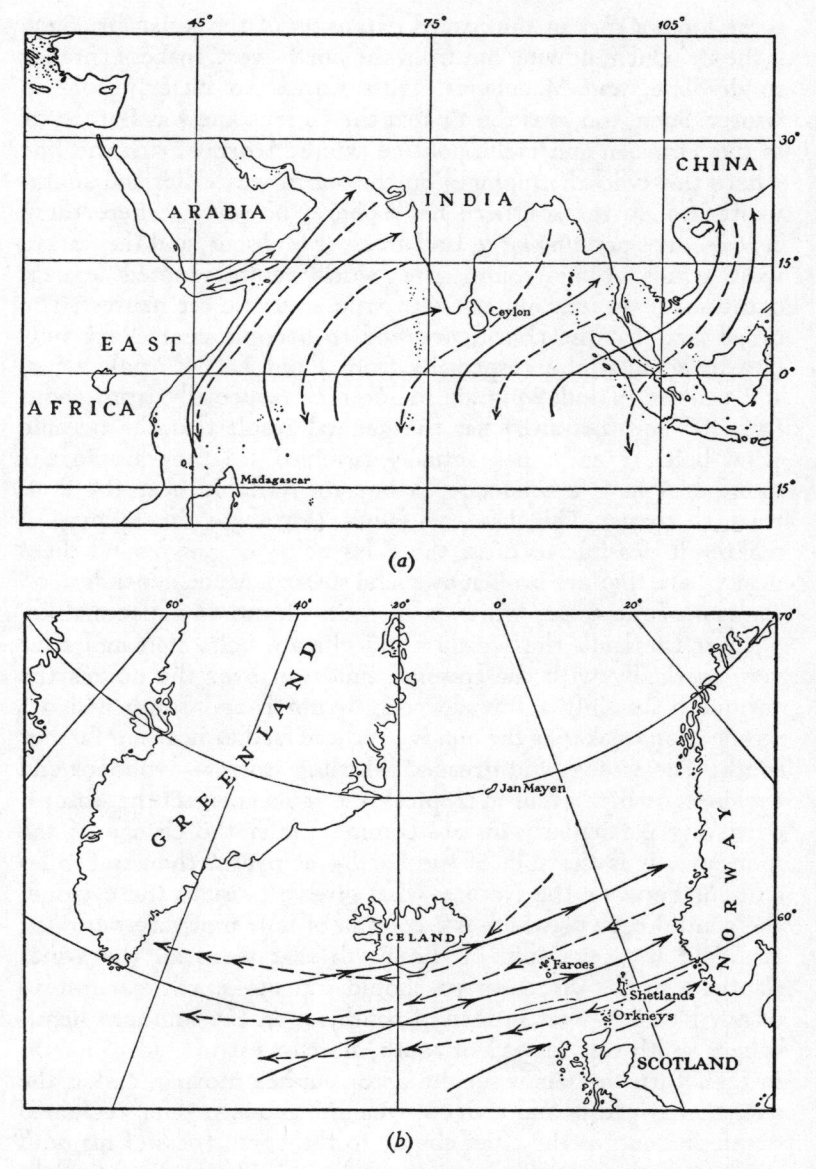

(a)

(b)

Fig. 7. Home blowing winds are associated with early ocean voyages in these areas.

these form a sort of southward extension of the polar air. Such is the air which, flowing out from the north-west, makes Labrador so desolate, and Manchuria, with Korea, so bitterly cold in winter. Such, too, was the air that the Greeks knew as Boreas, or as the Thracian and Hellespontine winds. Storms mark the line where this cold air impinges on the warm westerlies and south-westerlies. In the southern hemisphere, however, where there are no corresponding large land areas, it is absent, and the 'brave west winds' blow round the world uninterrupted except occasionally by encounters with true antarctic air masses. The broad fact that the continents tend to become centres for out-flowing air in winter (especially from about Lat. 40° polewards) and centres of inflowing air in summer (especially from about Lat. 40° equatorwards) has the general result that the oceanic wind belts, even if not actually reversed (as they are in the monsoons) have a tendency to become variable near the con-tinental coasts. This has the effect (fortunate for sailors) of making it possible to cross the calm belts, or pass round them near where they are broken by a land surface. Much depends upon the run of the coast, which is of course in no two places alike.

Over the lands the weather and climate belts shift and alter very markedly with the seasons, and even over the oceans the various belts shift a few degrees to north and south and get stronger or weaker as the Sun is overhead farther north or farther south. The violent and dreaded whirling storms—typhoons and cyclones—which occur in tropical waters also respect the seasons, occurring particularly in late summer or at the change of the monsoon. It is the spin of the Earth—at over a thousand miles an hour between the tropics—that gives a twist to the cyclone, the central core of which is a column of hot, moisture-saturated air. And it is the spin of the Earth that turns all the winds slantwise to the direction we should expect—north-east instead of north, south-west instead of south, or in the southern hemi-sphere south-east instead of south, north-west instead of north. In their turn the winds set the ocean surface moving, and as the ancient navigator, and indeed even the modern one, reckoned ocean currents as the chief enemy to the correctness of his dead reckoning, their pattern, too, must be considered. For it might be directly athwart the wind.

II

The Surface and Floor of the Sea

THE ancient Greeks correctly declared that the ocean was one, and washed the lands on all sides. And they argued this conclusion from the tides, which they found to rise and fall with the same rhythm in the Mediterranean Sea, the Indian Ocean and the Atlantic. Out in the open ocean the tide is scarcely noticeable; only in the shallow water round the lands is there an obvious flow of water to and fro. Twice a day the advancing tidal stream fills up the river estuaries damming back the fresh-water currents, and twice a day it runs back again, now reinforced by the natural outflow. Elsewhere it pours through straits and channels, first in this direction and then in that; it fills and empties gulfs, it divides and eddies around promontories, and piling up against obstacles it is reflected back from them. All this strong and confused motion results in the tidal races, tide-rips, overfalls and whirlpools, which are among the most significant 'dangers' to be marked and avoided by mariners in inshore waters. Homer tells the tale of how Odysseus barely escaped the horrible female Charybdis, the poet's personification of the insucking and outspewing tidal race through the narrow Strait of Sicily by Messina. It is true that he says that the phenomenon occurs thrice daily instead of twice, but the Greek writer Strabo, commenting on the story some eight centuries later, explains that this must have been a slip or a copyist's error. For he is at pains to prove Homer to have been a good geographer. But equally it may be imagined that the poet designed to heighten the horror. 'Charybdis', he says, 'swallowed up the salt water in terrible fashion. When she spouted, like a cauldron over a great fire she seethed up in a swirling mass, and the spray rose high in the air till it fell on the tops of the cliffs. When she swallowed up the salt water she showed deep down in her swirling whirlpool black sand at the bottom, and the rocks all round echoed a bellowing boom. Every man was pale with fear.'

Generally speaking, the rise and fall of the tide is small in the Mediterranean Sea, very nearly cut off as it is from the main ocean. As a consequence, the very much greater range between high and low water marks round about Britain occasioned great surprise to visitors from the Greek world. The first to come so far north was the Massilian astronomer Pytheas (*c.* 300 B.C.), whose writings are unfortunately lost, but according to Strabo (who considered him a liar) he reported tides of up to eighty cubits in the north of Britain. This, of course, is an exaggeration, but there are many ranges between thirty and forty feet, while at high water the waves may toss water up much further.

The first conception of the tides as great waves sweeping round an Earth completely covered with water has had to be abandoned in favour of a more realistic theory in which the tides in seas and oceans are composed of oscillations from end to end and side to side like those of water in a trough, but the time that elapses between the meridian passage of the Moon and high water is usually constant enough to allow a rough prediction of the tide. This time interval is the oddly named 'establishment of the port' which, as will be shown later, was early noticed by the pilots and sailors of the Atlantic coast. The requirements of modern shipping demand more accurate predictions than can be obtained by use of the 'establishment' or by the H.W.F. & C. (High water full and change) and L.W.F. & C. (Low water full and change) which replaced it, but in most places the relation of the time of tide to the Moon's meridional passage or to the age of the Moon gives a useful approximation.

The tidal disturbance produced by the Sun is less than half as great as that due to the Moon, but at new and full moon, when the two bodies are respectively in conjunction or in opposition as it is called, the two tides reinforce one another, producing the fortnightly spring tides. At spring tides the rise is higher and the ebb lower than the normal. On the other hand, midway between the dates of the spring tides, the Sun and Moon are in quadrature, that is to say are 90° apart, so that they pull against one another. Both the rise and the fall of the tide are damped down, producing the so-called neap tides, when flood and ebb are alike less marked. Further tidal inequalities are due to the fact that neither the orbit in which the Moon moves round the Earth, nor that in which the Earth moves round the Sun lies in the same plane as that in which the Earth rotates. This means that except at the equinoxes the lunar and solar tidal attractions are still somewhat at odds, even when the two heavenly bodies are in conjunction or opposition. It means, too, that the maximum upswing of the antipodal tide is not in the same latitude as that under the Moon, save on the two occasions each month when the Moon is in the plane of the equator. To all this must be added the fact that a strong, steady wind may either check the tidal flow or reinforce it, according to their mutual directions, a fact made sufficiently familiar by the occurrence of disastrous floods. It is not difficult to understand the dilemma in which the Breton lady found herself, when (according to Chaucer's tale), having promised to yield to her lover only when the rocks in the sea disappeared, she looked over the cliff to find they had actually done so. Fortunately (for she was happily married) the gentleman proved chivalrous, and did not exact the forfeit.

That the tides should be subject to the Moon, which was held to be the moist or watery planet, seemed quite natural to the ancients, and although the oldest surviving tide-table (one for London Bridge) dates only from the twelfth century, it is clear that a knowledge of the local tides must have been part of the equipment of fishermen and pilots since sailing began. Strabo relates that Aeolus, the legendary god of the winds, was in fact a seaman who was made king because he taught sailors how to navigate the dreaded Strait of Sicily in spite of Charybdis, a tidal stream. Ocean currents, however, are quite another matter. There is no chance of observing them day in, day out, as the tides are almost unconsciously observed by local coast-dwellers. Nor are they in themselves as regular and predictable as the tides. Little wonder, therefore, that they were always named by experienced sailors, and mentioned in old sea-manuals, as a major cause of faulty reckoning of the course made good. Both horizontal and vertical flows of water are necessarily set up whenever some local difference occurs in the density of the ocean water, and where a relatively constant factor is the cause of this disturbance of equilibrium, a more or less constant current will result. In regions of low rainfall and high evaporation, for example, such as the horse latitudes, the salinity and therefore the density of the water will be raised above normal. This denser water will sink and there will be a surface movement toward the area. The famous Sargasso Sea with its thick growth of the peculiar gulf-weed (looking as though you could walk on it, said an old sailor) and its seemingly stagnant, still waters, is a convergence centre of this kind. And as, like air movements, all streams of water are twisted slantwise by the Earth's rotation, the inward creep of water towards the Sargasso Sea takes on a clockwise direction, the entering rivulets (as they may be called) turning to the right as they leave the main current of water which in fact actually encircles this area.

This main current is to be attributed to the wind, for although there are many differences of temperature and salinity which cause movements, water freshened by equatorial rains for example, or by polar ice, the chief factor in surface movements is the wind. A steady wind may drive the water forward at two knots or more, but as the land areas interpose barriers in the path of a current and turn it aside, the water may eventually be moving athwart the wind, causing unsuspected leeway. The most important currents are those found in a broad belt on either side of the equator which sweep along from east to west, set up by the impulse of the trade winds. Some of the water thus set in motion eddies back between the two streams as an equatorial counter-current, and where, as happens in the Guinea Gulf and Indian seas, the trade wind is reversed in summer as a south-west monsoon, this counter-current becomes part of a powerful set from west to east. It was the strong west-flowing current of the trade-wind belt in the South Pacific Ocean that sent the raft *Kon-Tiki* bounding along from Peru to Polynesia. The corresponding current in the South Atlantic has a different history, for it encounters the out-thrust wedge-like shoulder of Brazil and is forced to divide into two. The southern branch turns south to circle around the horse latitudes, while the northern branch, skirting the north-east coast of the continent, is reinforced by the immense out-fall from the Amazon, and then itself unites with the west-flowing current set up by the north-east trades. A mighty body of water, therefore, approaches the West Indies, a large proportion of it entering and circling the Caribbean Sea, which lies in a deep basin. The only exit is between the peninsula of Florida and the Bahama Bank, and confined within this narrow channel, the current, now thoroughly heated, becomes both deep and swift. It is the famous Gulf Stream, with a velocity of 4–5 knots. Flowing northwards, the stream keeps a course parallel to the coast of the United States, but separated from it by a cool current of fresher water which is flowing in the opposite direction. The Gulf water is coloured a deep indigo blue, and its western edge is described as quite sharp, and easily distinguished from the greenish or greyish blue of the inshore current. By the time the Gulf Stream has reached Lat. 40°–45° the contrast between its waters and those nearer the coast has become dramatically marked, especially in the winter months, and this was a matter of moment in the old sailing-ship days. The American naval officer, Lieutenant H. Maury, writing a century ago about the approach to the New England and New York coasts, said: 'In making these parts of the coast vessels are frequently met by snow-storms and [north-west] gales which mock the seaman's strength and set at naught his skill. In a little while his bark becomes a mass of ice, and with her crew frosted and helpless she remains obedient only to her helm, and is kept away for the Gulf Stream. After a few hours' run she reaches its edge, and almost at the next bound passes from the midst of winter into a sea of summer heat. Now the ice disappears from her apparel: the sailor bathes his stiffened limbs in tepid waters; feeling himself invigorated and refreshed with the genial warmth about him. . . '

The bitter weather inshore is due not merely to the continental north-westerly winds. The Gulf Stream is now beginning to spread out fanwise and turn towards the east. Impinging upon its northern flank as it passes the edge of the Grand Bank is an immense body of outflowing polar water, laden in spring and early summer with melting pack ice, and with icebergs carved off from the glaciers of Greenland. One such destroyed the *Titanic* in April 1912. This cold current—the Labrador Current—thrusts icy fingers into the Gulf Stream and dips down beneath it, while the floating ice soon melts away. But when an air mass moves from above the surface of the warmer water and travels across the cold, its lower layers are chilled, and a fog forms over the Grand Banks. This region, including Newfoundland and the lower St. Lawrence, holds the world's record for the occurrence of summer sea-fog, for it is in summer that the contrasts of temperature are most striking. There is a matching fog area in the north-west of the Pacific Ocean, but it is less extensive, for that ocean does not, like the Atlantic, lie freely open to the polar basin. By the time it is crossing the Atlantic, now in the belt of prevailing westerly winds, the Gulf Stream has been slowed down to a drift rather than a current, but it is a warm drift and keeps the harbours of north-west Europe open and ice-free, its influence being felt as far north as Spitsbergen, while a branch turning off westwards in the Norway Sea prevents ice forming round the Faroes and along the southern shore of Iceland (Fig. 8a, b). The Gulf Stream drift creates in fact that 'gulf of winter warmth' that is responsible for the contrast, latitude for latitude, between the well-peopled lands on our side of the Atlantic and the desolation of Baffin Land and Labrador. As the warm water moves into the polar basin through the Norway Sea, so the cold ice-laden water moves out as a current through Denmark Strait, between Iceland and Greenland. Yet another current of polar water comes from Baffin Bay and through Davis Strait, keeping along the Canadian shore, and these two combine to feed the Labrador Current which crosses the Grand Bank. Strangely enough there is a warm undercurrent which is traceable along the west shore of Greenland, from Cape Farewell to Disko Island. Without it the Norsemen could never have colonized Greenland, for their farm settlements were along this shore.

The circulation of the North Pacific is but a pale copy of that of the North Atlantic. The current driven westward by the north-east trades encounters chains of islands or submarine ridges outside the monsoon region of the China Seas. A body of tropic water therefore turns northwards and, reaching about 4 knots east of Formosa, passes the south of the Japanese archipelago as the dark-coloured Kuro Siwo, in some sense a counterpart of the Gulf Stream. A drift of warm water crosses the ocean in the westerly wind belt and keeps the shores of British Columbia and Alaska open while the partly enclosed seas along the opposite coast of Siberia become icebound, for they are swept by icy north-west winds. In summer their still chilly green water is often hidden under a sea-fog. Nevertheless, like that on the Newfoundland Banks it is splendidly rich in fish.

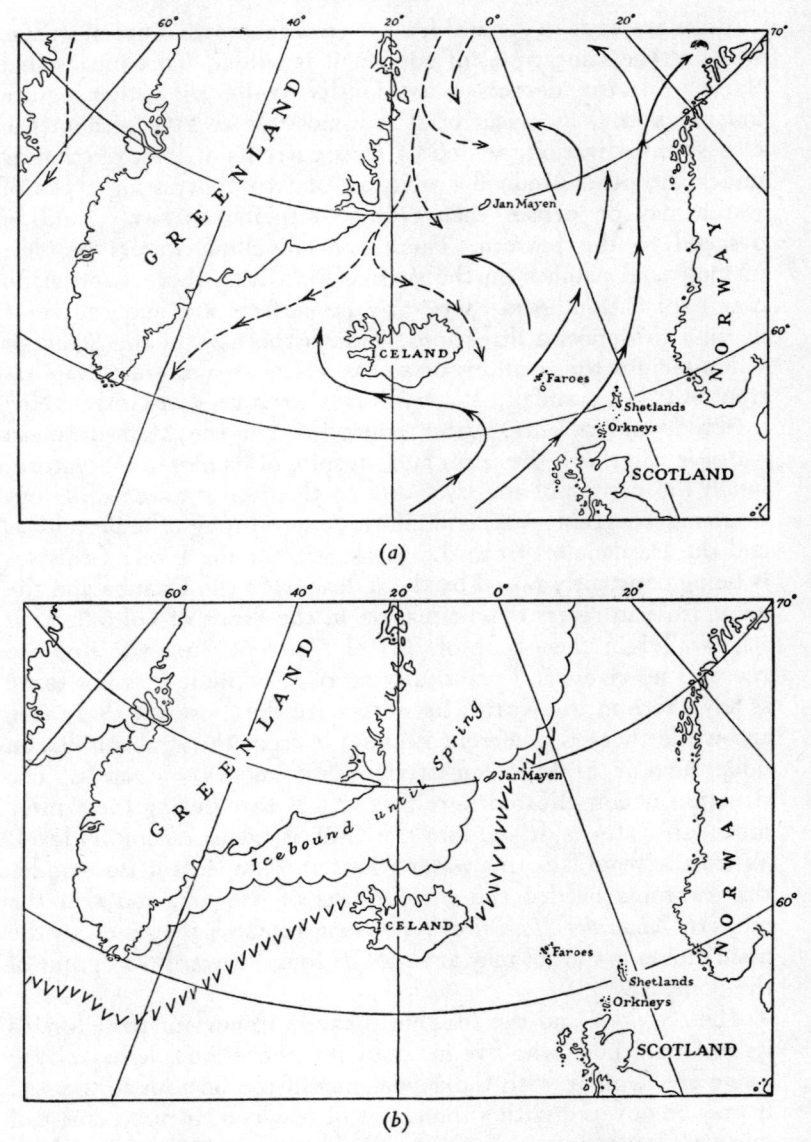

(a)

(b)

Fig. 8 (a, b). Warm water drifts polewards past Norway and cold water flows south past Greenland, so that the latter coast becomes icebound. The serrated line shows the occasional extension of the ice to Iceland.

Polar water is in general fresher than the normal ocean water, while at the same time, of course, it is colder. That means that the one factor decreases its density while the other, quite independently, increases it. It is impossible to say, without the closest investigation, which of two currents will flow over or under the other. Indeed a mixture of two contrasting types of water may be denser than either of them separately, and so descend to the bottom. There are upwelling waters besides, making cold patches on the surface for which there is often no easy explanation. And there may be surface and undercurrents flowing in opposite directions. This is the case in the Strait of Gibraltar, for the Mediterranean basin is an area of great evaporation but low rainfall. Its level is therefore constantly being lowered, and Atlantic water pours in. But the Mediterranean water is still much the salter and, in spite of its high temperature, much the denser of the two, and so this heavy water spills out as an undercurrent. A second inflow comes through the Bosphorus and the Dardanelles from the Black Sea, for the level of this sea is being constantly raised by the inflow from the Danube and the great Russian rivers. The situation in the Strait of Gibraltar has its parallel in the Strait of Bab el Mandeb, since the Red Sea receives no rivers and practically no rain, while it is as hot there as anywhere in the world. Its waters are the most salt there are, and while there is a current running in from the Arabian Sea, an undercurrent of this dense salt water moves outwards. The situation is complicated here by the fact that during the winter monsoon water is driven into the Gulf of Aden, raising its level. When the wind fails this water runs out again. It is little wonder that currents headed the ancient lists of dangers, and that the modern *Admiralty Manual of Navigation* remarks that even in the main currents the set may at times be found towards any point of the compass.

The sea-floor and the sea-shores are as important to sailors as its surface. Those who live in countries where the tide has a large range are familiar with the appearance of the bottom of the sea. It may be covered with a thick mat of seaweed, it may consist of fine or coarse sand, yellow or red, black or white, it may be a field of black mud, or display heaps of worn shells, pebbles or boulders. Again it may simply be a deeply creviced platform of rock. The theoretical explanation of this variety is simple enough. Rivers bring down fine mud which gradually settles down from the sea water. Breakers attack the cliffs, cutting a rock platform beneath them at the outer edge of which a terrace of the loose debris accumulates. But in point of fact the tidal streams, coupled with the varied structure and shape of the edge of the land, result in a complex picture that almost defies explanation. There are exposed stretches of shore and sheltered ones, they stand high or low, they are built of soft sands or clays, of soluble limestone, of hard granite. And the tidal streams sweep material alongshore, sorting it out according to size and weight, while the moving fragments grind and rub one another down. Moreover, marine plants and animals have their likes and dislikes as regards habitat, so that here they are present, there absent. The practical consequence of all this is that not only does the sea-floor show a pattern of heights and depths, of shallows and ravines, but also a pattern of floor-cover as variegated as the pattern of the fields, affording therefore sure 'landmarks' to the man who knows it. Broadly speaking it is a stable pattern, although a violent storm may disturb it in detail, but there are exceptional areas where changes must be expected, often from season to season. This is especially the case in tidal estuaries, where the banks and deep channels are always liable to shift, and in the deltas of rivers which come down in seasonal flood, here choking one arm with mud and there tearing out another. Even away from river mouths, wind and wave and tide may enter into all sorts of combinations to build up and to destroy, so that from time immemorial inshore waters have been 'pilotage waters', where local knowledge is called for. Nevertheless, at depths of five fathoms or more the floor-cover lies undisturbed, a slowly accumulating load of fine silt and organic residues.

All round the continents and larger islands there is a platform shelving out to a depth of about one hundred fathoms, after which there is a sudden drop into the true ocean basins, with depths of thousands of fathoms. This platform is termed the continental shelf, its outer edge the continental slope, and its width may vary from ten miles or so to some hundreds of miles. Indeed the whole of the British Isles, for example, stands upon the continental shelf of Europe. What the shelf is built of is a matter of debate. Is it a marine platform worn out by the waves, or a huge accumulation of debris from vanished continents? Very certainly in recent geological time—say within a hundred thousand years—it was dry land, and so is still scored with ancient river valleys and canyons. Parts of it were above sea-level within the last ten thousand years or less, as can be seen from the remains of submerged forests and peat beds exposed at low tide. But whatever the origin of the shelf, it has this character—the seaman once he reaches it is 'in soundings'.

It is on the continental shelf and slope, too, that the world's greatest fishing grounds are found, for the favourable conditions of depth and temperature, and penetration of light make them extraordinarily rich in marine life. The plants and animals that live on the sea-floor, and the helplessly floating larvae, animalculae and microscopic life which are all classed together as 'plankton', afford abundant food for larger creatures. Consequently the free-swimming fish and mammals (such as whales) come to the shelf to feed and spawn. It is an extraordinary evidence of the richness of arctic and antarctic sea-water in plankton that most whales can live entirely upon it, straining out enough to fill and refill their huge stomachs with 'crill' as the fishermen call it. The fishing banks off Newfoundland and New England, where the continental shelf is exceptionally wide, are among the most famous for their swarms of great fish which astonished their first discoverers. The Banks are a region too where the bottom dwelling or feeding organisms have a specially rich food supply, for the creatures constituting floating plankton carried from the tropics in the Gulf Stream and those carried from the arctic in the Labrador Current must die in their myriads where the waters meet, for they are very sensitive to change of temperature. A similar mingling of very different waters is to be found in many fishing grounds, on the Agulhas Bank, for example, and in the North Sea, which is seasonally penetrated by a tongue of the warmer, salter water from the Gulf Stream drift.

Off the coast of Peru there is a famous flow of cold water which comes north to replace the water driven westward by the trade winds. It is rich in plankton, and consequently rich in fish, which in turn attract myriads of sea-birds, whose droppings or *guano* on the rocky off-shore islands fertilized the fields which constituted the wealth of the ancient Incas. The guano is a carefully guarded state monopoly today, but its accumulation depends upon the absence of rainfall, which in turn depends on the trade winds blowing steadily away from the coast, and the off-shore waters being cold. About a quarter of a century ago there was a season during which the whole interlocking system temporarily collapsed. The trade winds for some reason weakened, and a stream of tropic water was driven south unopposed from the Gulf of Panama, so that the cold current failed and so did the fish, and the birds left, while torrential rain came down in what was normally a rainless area. This washed the guano off the islands, as well as flooding the land. There followed a return to 'normal', but such a sequence of events is a reminder that in any summary description of sea and shore there is a wide margin of error and over-simplification.

Once he is within sight of land (provided there is no fog) it is the shore profile rather than the sea-floor which claims the attention of the sailor. Promontories and inlets, capes and bays, have their twofold aspects of good and evil. A lofty promontory visible from a great distance gives promise of land and safe homecoming. Such was Cape St. Vincent, looming from perhaps fifty miles, the sacred cape of the ancients, marking the end of the habitable world. Yet in coastwise sailing a cape is difficult for a sailing ship to round, and seems to generate its own storms and gusts and contrary currents. Cape Malea at the southern extemity of Greece had an evil reputation, as appears more than once in the story of Odysseus as Homer told it. King Menelaus, sailing home from Troy, had his first misfortune at the 'sacred' Cape of Sunium, where Attica juts far out into the sea. Here his helmsman, 'the world's best steersman in a gale', was struck dead, oar in hand, by Phoebus Apollo. After burying him with due rites, the fleet sailed on and reached 'the steep bluff of Malea'. But 'Zeus who is always on the watch, took it into his head to give them a rough time, and sent them a howling gale, with waves as massive and high as mountains.' The fleet was scattered and Menelaus himself was driven right across to Egypt, while the rest of the fleet were wrecked on the southern shore of Crete, and their ships broken up. Here 'out of the misty sea there is a huge rock that falls abruptly to the water and the gales drive great rollers against a headland to the left'. It was Cape Leucas. Odysseus himself on another occasion was driven off course as he approached Malea by a combination of swell, current and northerly wind, and after nine days in the open sea found himself off the African coast, and approaching the land of the lotus-eaters. On every Mediterranean headland stood its shrine or temple to the gods, and the sailors' vows made to these were more often in supplication for safety than in gratitude.

A gulf or bay on the other hand is in the main beneficent, especially where a ship may be safely beached on shelving sand. Yet it carries the danger of embayment, and Odysseus very cautiously remained outside the fair-seeming, fjord-like bight of Laestrygonia. 'Here', he says, 'we found an excellent harbour, closed in on all sides by an unbroken ring of precipitous cliffs, with two bold headlands facing each other at the mouth, so as to leave only a narrow channel in between.' All the rest of the fleet steered in and tied up close together. But the 'nimble-witted Odysseus' made his ship fast to a rock at the end of one of the headlands, and when a murderous attack was made by the natives on those within 'I drew', so he continues his story, 'my sword from my hip, slashed through the hawser of my vessel, and yelled to the crew to dash in with their oars . . . with a sigh of relief we shot out to sea and left those frowning cliffs behind.'

PART TWO

NAVIGATION WITHOUT MAGNETIC COMPASS OR CHART

III

The Phoenicians and Greeks

THE oldest navigating instrument of which we have definite evidence is the familiar lead and line, which remains the safeguard of sailors to this day. The Greek historian Herodotus, writing five centuries before our era, mentions it quite casually when discussing the geography of Egypt, which he was the first to call 'the gift of the Nile'. 'On approaching it by sea', he says, referring no doubt to his own visit, 'when you are still a day's sail from the land, if you let down a sounding line you will bring up mud, and find yourself in 11 fathoms of water, which shows that the soil washed down by the stream extends that distance.' If the 'day's sail' is taken to be the distance sailed in the daylight hours, the ship would be about sixty miles from the shore, when the depth is somewhere about a hundred fathoms, so that no doubt the figure '11' is a corruption, arising from some long-ago faulty copying of the manuscript. But it is actually the case, according to the modern chart, that at this distance from the delta yellow Nile mud would be brought up on the lead. And the interesting point to notice is that at this early date it was already customary to put a lump of tallow at the bottom of the lead so that it brought up a specimen of the sea-floor deposit. This gave the experienced pilot additional and precious information about the distance and character of the shore that he was approaching, the mud or silt coming up in this particular instance indicating the nearness of a river's mouth.

To find himself 'in soundings', that is to say in depths of less than a hundred fathoms, was a first indication to the seaman that he was approaching land, although it might still, as in this case, be many hours' sail away. But if repeated casts of the lead showed a rapid shallowing, the inference was that land was close at hand. This was the case in the memorable story of the shipwreck of St. Paul. Owing to the continually overcast sky, it will be recalled, · they had completely lost their bearings. The north-easterly wind that drove them away from Crete had veered east and then south-east, and one night, about midnight, the mariners suspected (we are told) that land was near. What first aroused their suspicions does not appear. It was probably something the landsman narrator would not notice. Perhaps the noise of surf, the cry of sea-birds, or the bleating of a goat, perhaps some marked change in the rhythm or the violence of the waves. Whatever it may have been, the order was immediately given to cast the lead, and the suspicion became a certainty. The first sounding was 20 fathoms, quickly followed by one of 15 fathoms. Realizing that they were in imminent danger of being driven ashore, the sailors put out four stern anchors and the whole ship's company waited fearfully for the dawn. 'And when it was day, they knew not the land', or in other words the pilots recognized no landmarks, and the shore profile was unfamiliar. The ship was beached, and they found themselves in Malta.

As the sounding-line was obviously no novelty in Herodotus' day, it is reasonable to suppose that it had long been part of the equipment of the famous Phoenician sailors whose voyages took them as far as the English Channel. It was probably used, too, by their predecessors, the Cretans, who sailed over the whole Mediterranean Sea, by night as well as by day. The unit of measurement, the fathom, was the length of a man's outstretched arms, and the line may have been knotted at this interval, for an ancient Egyptian wall-painting ,which dates back to the second millennium B.C., shows field-surveyors using a knotted cord, which is being unwound from a neat flat coil. Architects and builders also used the plumb-line, an implement closely analogous to the sounding-line, for measuring heights as well as for deter-mining the vertical. It is mentioned, for example, in the eighth century B.C., by the prophet Amos, who says: 'Behold the Lord stood on a wall made by a plumb-line, with a plumb-line in His hand.' And again a century and a half later Ezekiel, in describing the survey of a visionary Holy City, writes: 'There was a man . . . with a line of flax in his hand, and a measuring reed.' This reed, he tells us, had a length of six great cubits, or of nine feet. Just such a long rod or pole is seen in the hands of the pilot of one of the ancient Egyptian sailing ships shown setting out for the Land of Punt, which lay in or perhaps beyond the Gulf of Aden (see

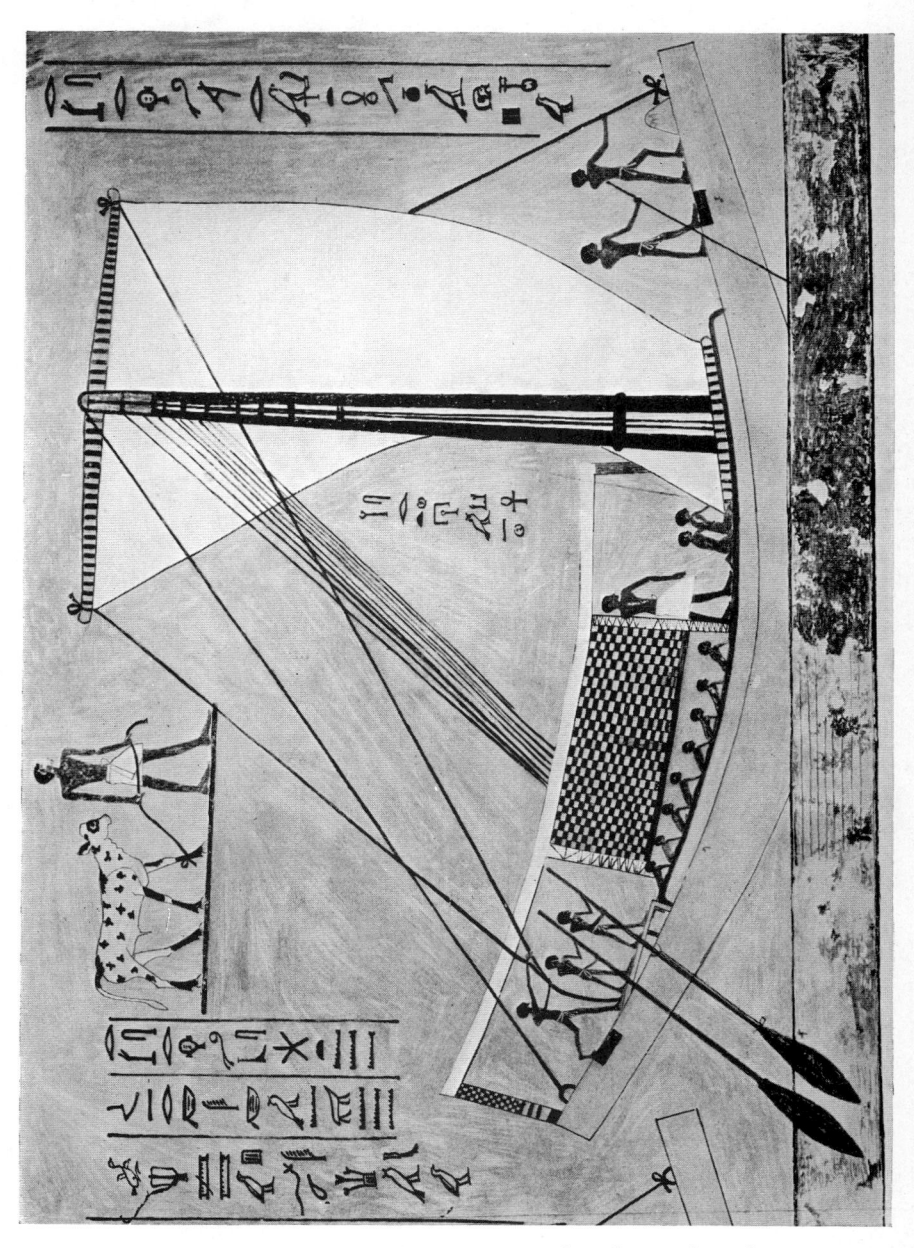

II. In this Egyptian sailing ship (before 1500 B.C.) the pilot stands in the prow with his sounding-rod while a sailor behind the two steersmen holds the braces. One of the steering oars is used through a thole pin on the starboard thwart, the other can be transferred to port. The bifid mast is braced fore and aft.

III. The eight personified winds are sculptured on the faces of the octagonal Tower of the Winds in Athens. These eight 'quarters' or winds were halved and halved again to give thirty-two 'points' or rhumbs.

also Plate II). He is directing the helmsman by hand signals from the prow. As the ship left harbour in the Red Sea an infinity of islets, rocks, reefs and pinnacles of coral would be encountered, and the pole was no doubt used to probe the intricacies of the channel to be followed. It is doubtful whether the ship tied up at night, although this was the Arab practice in the seventeenth century, and a source of great annoyance to European travellers. They too report the conning of the ship from the prow, and indeed it can be met with in small dhows today. It is interesting to recall that Odysseus also carried a long pole aboard. He seized and used it to quant his ship away from the shore when it was in danger of being swept within reach of the infuriated Cyclops whom the hero had blinded. There is no mention, however, of his taking soundings.

The pilot of these early days, as well as being skilful in taking and interpreting depths, was undoubtedly familiar with the feel, character and behaviour of the winds from different quarters, besides having his memory stored with landmarks and coastal profiles as they appeared from the sea. 'Wind', as has already been pointed out, meant practically the same thing as 'direction', and a wind was often named from the country from which it blew. But it was necessary for the seaman to know also the country to which it would carry him, and this he learned from the accumulated experience of fishermen and sailors in his locality. Thus Pliny (who had been a naval officer) when he wanted to define a local French wind, called Circius, said it was one that would carry you from Narbonne across the Ligurian Sea to Ostia, that is to say a north-north-west wind. Or again, he used the expression, 'from Carpathus is 50 miles with Africus to Rhodes', meaning that Rhodes was 50 miles east-north-east of Carpathus. His phrases appear to have been borrowed from current sailing directions, of which more will be said presently, and they were obviously practical and sensible. A shipmaster wishing to sail from Carpathus to Rhodes had to wait until Africus was blowing. The following wind, or wind on the poop, generally termed the 'fair' or the 'prosperous' wind was all-important in the very earliest days of sailing, when a loose billowing square sail, which was not capable of much manipulation, was used. It was soon torn to rags, besides, by a really strong wind, so that in bad weather it was hastily lowered and the men took to the oars. It was because the winds in the Red Sea are seasonable and reliable as a rule both in direction and character that long sailing voyages could be undertaken there at so early a date. For the same reason there was very ancient sailing with the monsoons between India and Africa, but of this, unfortunately, we know very little. But the sense of security bred where a return wind unfailingly recurs can easily be imagined.

The northerly winds, or Etesian winds, which blow in the Levant and over Egypt in the summer gradually extend southwards as the monsoon develops over India, when they become canalized by the high rocky shores of the Red Sea, so that they blow throughout its length as north-north-west winds. As the monsoon changes they weaken and retreat, while a south-south-easterly blows in from the Strait of Bab el Mandeb, and gradually extends northwards. It may even occasionally reach the Gulf of Suez. Outside the Strait the trend of the coast of Arabia, like that of the 'Horn' of Africa, is roughly parallel to the alternating south-west and north-east monsoons, thus inviting coast-wise travel, and although the land of Ophir has never been identified, there is no reason to doubt that the famous fleet, equipped by King Solomon nearly three thousand years ago, went out into the Gulf of Aden. In the northerly part of the Red Sea, however, the southerly wind cannot be relied upon, and squalls and storms are common and often violent. The tides, too, which rise and fall in the Gulfs of Suez and Akaba set up dangerous cross-currents at the extremity of the Sinai peninsula. Navigation here is consequently difficult, and the more important Egyptian ports, Kosseir and Berenice, were placed well to the south, so as to catch the return winds. In Roman times, however, the more northerly Myos Hormos was used, as well as Leuce Kome, a little farther south on the Arabian shore. Traffic for these Egyptian ports had to move overland from the Nile waterway, and it was probably because of the less favourable sailing conditions in the Gulfs that the ancient canals built to connect the river and the sea there were never a great success. And when Jehoshaphat built a 'Tarshish fleet', presumably a fleet of long-distance merchant ships, to take the Ophir voyage once more, they were broken up at the very start in the Gulf of Akaba.

Fig. 9. Personified winds depicted in Peter Apian's *Cosmographia* (1551).

The Red Sea sailings, however, must have played an important part in advancing knowledge of astronomy, as well as in improving the art of navigation. For they involved a change of latitude, and so of Sun and star, of at least 20°, to which must be added the 5° or 6° sailed northwards by Egyptian mariners, who are known to have trafficked with Byblos in Syria from very remote times. The summer sky is brilliant and cloudless, and the pilot watching it at night could not fail to notice that the Great Bear, who in home waters never bathed in the sea, began to dip her limbs below the horizon as his ship came towards the lands of frankincense and cinnamon. He would note, too, the growing splendour of Canopus, a brilliant star barely glimpsed in Phoenicia. 'These phenomena', wrote Pliny when discussing the shape of the Earth, 'are most clearly discovered by the voyages of those at sea . . . the stars that were hidden behind the curve of the ball suddenly become visible, as it were rising out of the sea.' The sea, in fact, offers the clear-cut horizon which is usually wanting on land, while it is the sailor's business, besides, constantly to scan this horizon.

The first seaman of whom we are actually told that he steered by the stars is Odysseus. And while it is true that he was merely the mythical hero of an epic poem, we may be sure that his exploits embody the traditional method of seafaring of a still older day than Homer's. When leaving Calypso's island with a happy heart, although all his companions were gone, he 'spread his sail to catch the wind and used his seamanship to keep the boat straight with the steering-oar. There he sat and never closed his eyes in sleep, but kept them on the Pleiades, or watched the late-setting Boötes [Arcturus] and the Great Bear. . . . It was this constellation that the wise Goddess Calypso had told him to keep on his left hand as he made across the sea. So for seventeen days he sailed upon his course.' By keeping the always visible Bear on his left, Odysseus was of course sailing east, and since the Pleiades and Arcturus differed nearly eleven hours in right ascension one or the other would also always be visible in the night sky, and sometimes both. The Pleiades, so Pliny tells us, rose with the Sun six days before the Ides of May, so that by the time autumn came they could be visible all night, whereas Arcturus, who was first seen forty days before the autumnal equinox, would be high in the sky when the sailing season opened in spring. In Homer's times this was a zenith star in the mid-Mediterranean Sea, and was termed 'late-setting' or 'slow-setting' because it only remained below the horizon for six hours out of the twenty-four. Watching these familiar stars as they neared and crossed the meridian Odysseus would steer due east. And on the eighteenth day of his voyage he picked up a familiar landmark. 'There hove into sight the shadowy mountains of the Phaeacian country, which jutted out to meet him here. The land looked like a shield laid on the misty sea.' Perhaps it was Corsica, for Odysseus had been adventuring in the far west of the world. But at this juncture the enraged Poseidon, returning from a distant journey, observed the hero's happy fortune, and proceeded to break up his ship by a succession of squalls from every quarter. Odysseus was left clinging to a spar, but luckily the goddess Athene now intervened. Stilling the squalls, she sent out a northerly breeze, on which he paddled forward for two days and two nights until, lifted high on the crest of a great wave he once more caught sight of the Phaeacian shore. But, says Homer, 'great seas were battering the iron-bound land, and all was veiled in spray'. After a vain attempt to drag himself on to a rock Odysseus had the happy notion of swimming along the coast outside the breakers in search of some natural harbour. He soon felt an outflowing current which told him he was off a river mouth, and on his offering a prayer to this unknown river it graciously checked its current (perhaps the tide turned) and he was able to swim in and land.

Although all this was but part of a story, it was a story told to seafaring listeners, and told by a poet who knew the ways of the sea and of sailors. It is in the vivid, selective detail which Homer gives about harbours and landing places that we can trace the beginnings of the port-books and pilot-books of later ages. Here, for example, is the way he describes a safe harbour on the island which lay off the Cyclops' country. 'You need neither cast anchor nor make fast with hawsers: all your crew have to do is to beach their boat and wait till the spirit moves them and the right wind blows. Finally, at the head of the harbour there is a stream of fresh water running out of a cave.' Then there is the harbour called Phorcys, in Odysseus' homeland, Ithaca, to which the

Phaeacian sailors, 'knowing the spot', brought him in the early morning when lovely Venus was rising and ushering in the dawn. This harbour had 'two bold headlands squatting at its mouth so as to protect it from the heavy swell raised by rough weather in the open, and allow large ships to ride inside without so much as tying up, once within mooring distance of the shore'. At the head of the inlet, too, there were 'springs that never fail'. Whether or no a harbour had an easily recognizable entry, provided a safe berth and had a good water supply—these were the things that a pilot needed to know. In Phorcys harbour the Phaeacians drove the ship up on the beach with their oars, deposited the sleeping Odysseus upon the sand, together with all his gear, and immediately sailed away.

Homer presents the Phaeacians as being the finest seamen in the world of their day. They cannot, it is true, be identified with any actual people, yet they have much in common with the ancient Cretans as these are described by modern archaeologists. Like the Cretans they were renowned, for example, for their skilful and abundant bronze work, and their splendid textiles, as well as for their seamanship. Their great city was a seaport, built on an island connected with the mainland by a causeway, as Tyre was to be centuries later. The port had twin harbours, as was not uncommon in the Mediterranean, so that one or other could be entered whatever the wind. Lining these harbours were the slipways of individual shipowners, while between them lay a magnificent public square built of massive stone, on which stood a temple to Poseidon. A supreme nautical skill, based no doubt on superior techniques which they kept secret, appears to have given rise to a popular belief that there was some magic about the Phaeacian black ships. This myth Homer chooses to put into the mouth of their king, who tells Odysseus that their ships 'have no steersmen, nor steering-oars such as other craft possess'. 'Our ships', he said, 'know by instinct what their crews are thinking, and propose to do. They know every city, every fertile land, and hidden in mist and cloud they make their swift passage over the sea's immensities with no fear of damage and no thought of wreck.' It might almost be thought that the Phaeacians were already using rudders and had the magnetic compass, although this of course is out of the question. Less improbable is the surmise that the myth arose because the ships were decorated at the prow with the *oculus*, the eye of the God, still to be seen on many Mediterranean vessels, and which, so simple folk believed, helped a vessel to find its own way in safety.

In Homer's own day, which was long after the destruction of the Cretan sea-power, the most able and active seafarers known to history were the Phoenicians, who also appear in his story, although often in no very favourable light. Among them the men of Sidon were reckoned the most highly skilled, Sidonian pilots being selected for difficult or important enterprises. Greek writers speak of the Phoenicians as their masters in maritime affairs, and in particular as teaching them the better way to distinguish the north by means of the Lesser Bear in place of the great Wain or Plough. As Strabo puts it, the smaller constellation 'did not become known as such to the Greeks until the Phoenicians so designated it for the purpose of navigation'. And here he is echoing the poet Aratus, who in describing the heavens in his *Phaenomena*, composed about 276–274 B.C., writes as follows: 'On either side the axis ends in two poles: but thereof one is not seen, whereas the other faces us in the north, high above the ocean. Encompassing it two bears wheel together. . . . One men call by the name Cynosura, and the other Helice. It is by Helice that the Achaeans [i.e. Greeks] on the sea divine which way to steer their ships, but in the other the Phoenicians put their trust when they cross the sea. But Helice appearing large at earliest night is bright and easy to mark, while the other is smaller, yet better for sailors, for in a smaller orbit wheel all her stars. By her guidance, then, the men of Sidon steer the straightest course.' The story can be traced back still farther, for Aratus was here putting into metre an earlier astronomical work. This was the prose description by Eudoxus of Cnidus, his predecessor by a hundred years, and the man who has been termed the greatest mathematician and astronomer of all time.

It is from Eudoxus' day that the Greeks can be spoken of as good astronomers, and the first sailor-astronomer of whom we have definite knowledge, the famous Pytheas of Massilia, who has already been mentioned, lived only two or three generations later. Although Pytheas' own writings are lost, and we know of his work and travels only through fragmentary references and quotations by later writers, including Strabo who vilified him as an 'arch-liar', we have evidence of his exceptional knowledge and precision of observation from another very famous man, the astronomer Hipparchus. Hipparchus made the first star-list, and wrote a commentary on the very work of Aratus which has just been quoted. And it is here that he censures Eudoxus for speaking as if there were a fixed Pole Star. This is what he says: 'Eudoxus displays ignorance about the North Pole in the following passage: "There is a star which remains always motionless. This star is the Pole of the world." In fact there is no star at the pole, but an empty space, close to which are three stars which taken together with the point of the pole make a rough quadrangle, as Pytheas of Massilia tells us.' Which exactly were the three stars used by Pytheas to find the precise celestial pole is a matter for discussion, but he was obviously not satisfied to use Kochab, over 7° away; and indeed it is possible that Eudoxus was referring to one of several inconspicuous stars that were much nearer the axis of rotation than Kochab, or to a hypothetical 'axis' star.

It is necessary of course to distinguish between the sailor's use of Sun and star to divide the circle of the horizon—merely to orient himself that is to say—and the use of the heavenly bodies to find position on the Earth's surface, in other words the latitude. The one demanded simple observation, while the other involved measurement. It required, besides, logical thinking behind the measurement, and the capacity to compare measurements, and to understand their significance. Pytheas fulfilled all these intellectual requirements, and was remembered for his expert use of the instrument called the gnomon, by means of which he made a very close determination of the latitude of his native Marseilles. The gnomon was a very accurately set vertical pillar standing on a horizontal base across which its shadow fell. The exact bisection of the angle between two shadows of equal length was sufficient to establish a precise meridian line, and so to divide the compass or horizon circle. Furthermore the ratio of the noon shadow-length at the equinox to the height of the column is what today would be called the cotangent of the latitude, and it was this ratio which Pytheas obtained with extreme care, giving the latitude of Marseilles as 43° 3', correct within 15'. Such an instrument is not, of course, portable, and on his famous Atlantic voyage to the Arctic Pytheas made enquiries into two other phenomena which are functions of latitude, namely the lengths of the longest and shortest days and nights, and the local 'arctic circles', as defined by the extreme circumpolar star. The limit of his journey, the island of Thule, was long accepted by later geographers as the limit of the habitable regions of the Earth. In Thule he found (says Strabo, quoting from Posidonius) that 'the summer solstice was the arctic circle', meaning that no star more than 24° from the equinoctial circle ever set below the horizon. Or in other words the pole was elevated 90° – 24° or 66°, so that what we today term the arctic circle passed, so he considered, through Thule. Strabo thought it was nonsense to say that there were men so far north, for he considered that even Ireland was barely habitable, whereas Pytheas had said that Thule was six days' sail (about 600 miles) north of Britain and near the frozen sea, the edge of which he said he had actually seen.

Of his further observations we have a record in the *Elements of Astronomy* written by a Greek named Geminus about 70 B.C., who quotes from a work written by Pytheas entitled 'Of the Ocean'. 'The barbarians [i.e. natives] showed us where the Sun goes to rest. For it was found that in those regions the night was quite short, consisting in some places of two hours, in others of three, so that only a short interval elapsed between the setting of the Sun until it rose again.' All this is consistent with Pytheas' ship having approached south-east Iceland, which lies in Lats. N 63° to 64°, the 'barbarians' being people who either lived there or visited it for the fishing. But the ever-critical Strabo declared that Pytheas had merely used his knowledge of arithmetic and astronomy to make calculations which would bolster up his pretence of having visited the arctic. Yet the Massilian's description of the 'curdled sea'—the edge of the ice-sheet—which he says he actually saw, carries conviction. He called it a 'sea-lung', a mixture of the elements of earth, water and air where one could neither walk nor sail. And the rhythmically heaving surface of mingled brash ice and sea-water, glimpsed through the wet, grey mist which so often hangs over it in summer, might well have suggested to a Greek observer the breathing of the fabled Demogorgon hidden beneath the surface.

Pytheas' Atlantic voyage, which took place in the days of Aristotle, was an isolated event so far as the Greeks were concerned, for the Phoenicians, who had colonized the western Mediterranean and the adjoining Atlantic coasts, tried to keep the oceanic trade to themselves. As far back as 600 B.C. they were sailing to Cornwall for tin and had observed that the people of Brittany voyaged to a Holy Isle, i.e. to Ireland, in their skin boats. Southward, too, at about the same time, Hanno led his famous voyage as far as Sierra Leone, of which a record was set up at Carthage. But the record contains no hint of navigational methods, or of any astronomical observations. It was, however, from the daily experiences of sailors that the earliest arguments were provided about the global shape of the Earth: ships disappear hull down, a lantern must be placed at the masthead to be visible at a distance, while from the masthead the lookout saw land that was invisible from the deck. As one sailed north, besides, more of the stars became circumpolar ones, while as one sailed south new stars rose, culminated and set, and at length the tail tip of the Lesser Bear touched the horizon.

All this was to become the commonplace of the academic text-books, yet the first direct evidence we have that the height of the stars was used by sailors to find their position comes very late. It is to be found in the epic poem written by Lucan, a young Roman gentleman at Nero's court, which may be dated about A.D. 63–65. The poet describes the defeat of Pompey by Caesar at Pharsalus, nearly a century before his own day, and he makes the fleeing Pompey interrogate the sailor who is to carry him to a port in Syria. The fugitive general asks:

> 'Rectoremque ratis, de cunctis consulit astris,
> Unde notet terras, quae sit mensura secandi
> Aequoris in caelo, Syriam quo sidere servet,
> Aut quotus in Plaustro Libyam bene dirigat Ignis?'

which may be rendered: 'He asks the master of the ship, how does he know the lands by all the stars he observes? What is his guiding point for cleaving the sea? Which star serves for Syria, or which fiery spark in the Plough shows him the right way to Libya?'

And to this the sailor replies:

'Signifero quaecumque fluunt labentia caelo
Numquam stante polo miseros fallentia nautas
Sidera non sequimur: sed qui non mergitur undis
Axis inocciduus gemina clarissimus arcto.
Ille regit puppes. Hic cum mihi semper in altum
Surget, et instabit summis minor ursa ceruchis,
Bosporon, et Scythiae curvantem litora Pontum
Spectamus. Quidquid descendit ab arbore summa
Arctophylax propriorque mari Cynosura feretur,
In Siriae portus tendet ratis?'

which is as much as to say: 'We do not follow any of the restless stars which move in the sky, for they deceive poor sailors. We follow no star but one, that does not dip into the waves, the never-setting Axis, brightest star in the twin Bears. This it is that guides the ships. When the Lesser Bear rises and stands high above me in the yards, we are looking towards the Bosphorus, and the Pontus washing the shore of Scythia. The more Arcturus comes down from the mast-top, and the nearer Cynosure is to the sea, then the ship is approaching a Syrian port.'

That for Arctophylax we may read Arcturus is clear from what Aratus says in his *Phaenomena*: 'Behind Helice like one who drives, is borne along Arctophylax, whom men also call Boötes since he seems to lay hands on the Wain-like Bear . . . beneath his belt wheels a star bright beyond the others, Arcturus himself.'

So too, although Cynosura was one of the names for the whole constellation of the Lesser Bear, it meant also the Pole Star or axis, the modern Kochab. The actual pole would sink 12° nearer the sea in the course of a voyage from the Pontic or Cimmerian Bosphorus (the strait into the Sea of Azov) to the port of Tyre, but it is a real difficulty that the star Kochab, then nearly 9° from the pole, could still be spoken of as fixed. Did the pilot habitually observe its upper culmination, or its position only when it lay east–west with its 'brother', the hinder 'guard' of later days?

And there is a second difficulty in this case about Arcturus. A zenith star in about Lat. 36° in Homer's day, the precession of the equinoxes had by the opening of the Christian era made it a zenith star in Alexandria (Fig. 5). Throughout the voyage southward to Syria, instead of descending from the masthead, it would have risen higher and higher towards it. To a poet, however, such details are immaterial; a star must rise or sink to fit his verse. And the important point for our purpose is the indication that it was customary to measure star altitudes against the mast and spars. This told a sailor who had lost his reckoning whether he was north or south of his home port, or of some other desired haven for which he knew the stars. It is probable that a man who aspired to be master or pilot of a ship had a far keener eye for the stars than we can find evidence for in written history. At one point in his story Homer, for example, says: 'In the third watch of the night, when the stars had passed their zenith, Zeus the cloud-gatherer sent us a gale.' And in another place: 'In the third watch of the night, when the stars had turned their course.' Now at all times of the night some star or another is reaching and passing its zenith, so that the reference must be to the particular 'clock stars' that had been picked out for time-telling in Egypt since about 2500 B.C. One such 'decan' star (as they were called) crossed the meridian about every forty minutes, and every ten days a new one appeared and joined the procession at dawn, while an old one disappeared about sunset. The reason for this, of course, was the fact that star time gets nearly four minutes fast a day, but for ten days a particular star's meridian crossing was deemed 'near enough' to midnight. The third watch would thus be indicated by the fact that the current decan star and those in front of it were descending, having 'turned their course'.

In Lucan's poem Pompey was fleeing to Syria, and it was perhaps a Phoenician sailor who answered him. It is tempting to believe that the Phoenicians had long since made use of star altitudes on their Atlantic voyages. Strabo, who had learned philosophy under a Sidonian master, tells us that the Sidonian people specially studied arithmetic and astronomy, which they needed, the one for their trading accounts and the other for their night-sailing. It will be remembered that the prophet Ezekiel in his tirades against Tyre uttered the sentence 'Thy wise men that were within were thy pilots', which seems to suggest that the most learned among them were chosen for this office. If so, it was only natural that when the Greeks tried to challenge the Phoenicians in the Atlantic a man of Pytheas' calibre had to be aboard.

The ancients, we must always remind ourselves, were satisfied with very rough measurements. Among astronomers the division of the circle into degrees, minutes and seconds was introduced by Hipparchus, but laymen continued to measure the elevation of the Sun and stars in cubits. Strabo, for instance, when comparing the latitudes of different regions says: 'On winter days there, the Sun ascends only 4 cubits among the people, who are distant from Massilia 9100 stadia [or 12° latitude].' The cubit is reckoned to have been equal to two degrees of angular measure. The early Egyptian astronomers had been satisfied to measure the height of the stars against a figure seated at a certain distance in front of them—the heavenly body stood at his shoulder, at his ear, or at the crown of his head. Measurements, too, of altitudes and distances in terms of the width of the finger or of the palm held up at arm's length are also very ancient. A man's middle finger would cover roughly two degrees, a palm or fist eight degrees. The sailor, looking up from the fixed steersman's seat in the stern might divide his mast into hand's breadths, and so watch the stars. But whether this is what he did we cannot know. Following Alexander's conquests, when Babylonian astronomy became the possession of the Greeks, such instruments as the dioptra, the armillary sphere, the scaphe or *polos*, and many others, came into use. But this was among the mathematicians only, and there is no suggestion that they were used aboard ship, for which indeed they were quite unsuited.

By the fourth century B.C. there appears to have been sufficient literacy among the masters and pilots of trading ships and naval vessels to warrant the compilation of written sailing directions. Or perhaps it would be truer to say there was sufficient literacy to bring port-books and pilot-books into more general use, for the earliest of such works to survive seems to be derived from yet older sources. This document is the *Periplous of Scylax of Caryanda*, so called after a famous sea-captain of an earlier age who was remembered for the voyage he made in about 510 B.C. from the Indus to the Red Sea. It was undertaken at the instance of the Persian King Darius I, to whom Scylax brought a detailed description of the coast. The *Periplous* itself, however, says nothing of this, for it is confined to an account of the harbours and land-marks of the Mediterranean Sea and the adjacent parts of the Atlantic. We are afforded a glimpse, however, of how material for such a document had gradually been collected in a story told by Herodotus, who was writing about 450 B.C.

This same King Darius had planned an expedition against Greece, and as a preliminary step ordered an examination of its coasts. As, however, the Persians were not a sea-people he hired Phoenician ships for the task, and obtained skilled personnel from Sidon. His consort's Greek physician, who was a native of Magna Graecia (southern Italy), was also added to the ship's company. The little fleet, consisting of two triremes and a laden round-built merchant ship, sailed directly to Greece from Phoenicia, and 'when they had made the land [says Herodotus] they kept along the shore and examined it, taking notes of all that they saw, and in this way they explored the greater portion of the country, and all the most famous regions, until at last they reached Tarentum.' Here the Greek doctor found himself near his home town, and turned traitor. The trading vessel was seized and after this loss the Phoenicians made for home, but on the way they were wrecked on Cape Leuca, the extreme headland of the heel of Italy, a point always difficult to double (Fig. 1). An interesting sidelight is cast on the administrative practice of those days by the fact that when the three foreign ships first arrived at Tarentum the local authority immobilized them by confiscating their steering gear, which was subsequently re-turned.

Quite apart from such special surveys as those ordered by Darius, the fact that there were Phoenician and Greek colonies and trading stations throughout the Mediterranean and Black Seas, and even in the neighbouring parts of the Atlantic, with active intercourse between them all, made it a simple, if labor-ious, matter to collect the pilots' lore into a comprehensive 'periplous' or circumnavigation of the whole area. The *Periplous of Scylax of Caryanda* commences at the mouth of the Nile and proceeds westwards along the Libyan coast. One or two quota-tions will show its style and the sort of information given: 'The mouth of the Bay of Plinthine [the modern Arabs Gulf] to Leuce Acte [Ras el Kawais] is a day and night's sail; but sailing round the head of the bay is twice as long. From Leuce Acte to the harbour of Taodamantium is a half-day's sail. . . .' Or again, after many details have been given of the north-west African coast, they are summarized as follows: 'Coasting from the Pillars of Hercules to Cape Hermaea is two days. From Cape Hermaea to Cape Soloeis coasting is three days, and from Cape Soloeis to Cerne seven days' coasting. This whole coasting from the Pillars to Cerne Island takes twelve days. The parts beyond the Isle of Cerne are no longer navigable, because of shoals, mud and seaweed. The traders here are Phoenicians.'

Except for very short stages which are given in stades the unit of distance in this early pilot-book is the day's sail. This does not mean the actual time taken; it is the theoretical distance which a normal ship would accomplish during a twenty-four-hour run with a fresh following wind. According to the *Periplous* the pseudo-Scylax reckoned this as 1000 stades, which may be read as 100 or 125 Roman miles according to the value given to the stade. Another authority states that a ship averages 70 miles during the daylight hours and 60 during the night, making a day's sail of 130 miles, but obviously the unit cannot have been an exact one. The geographer Eratosthenes, who measured the distance between Rhodes and Alexandria astronomically as $5\frac{1}{2}°$, made it 3850 stades in length, for he reckoned (from other measurements) a degree as 700 stades. But, says Strabo, some seamen gave it as 4000 stades, others as much as 5000; the wide discrepancy suggesting the uncertainty of the units employed. It is Strabo who tells us, too, that the astronomer Hipparchus (150 B.C.) when mapping the known world 'trusted to the sailors' for the measurements of the western Mediterranean basin, about which he himself was rather dubious. We know, in fact, that Hipparchus used periploi, as Strabo did himself in his detailed geographical descriptions. Herodotus, when he wished to record dimensions, e.g. of the Black Sea, gave them in terms of the day's sail, but in respect of the Caspian Sea he specifically used the day's voyage in an oared boat. No doubt the native sailors on the Caspian at that day did not use sails, for it seems likely that the Greek traders settled on the Russian coast, of whom Herodotus had made his enquiries, had personal know-ledge of the inland sea. The 'day's voyage' in this case appears to have been about forty miles. The 'day's journey' was the land unit corresponding to the 'day's sail', and Herodotus points out that this would differ in length according to the type of traveller —the ordinary man, the lightly equipped soldier, an army unit. Little more than a century after his day, however, Alexander the Great was employing surveyors or 'bematists' to step out the roads in his newly conquered empire. A precise unit of length of 1000 paces, or one mile, had thus already been established, but it still remained true that most distance measurements in itineraries and geographical descriptions were merely translations of the hypothetical 'day's journey' and were not derived from surveys. How the accepted equivalent in stades or miles of a 'day's sail' was arrived at we do not know. But ever since sailing began masters and pilots have always prided themselves on knowing the 'feel' of their ship and how much way she was making, and during coastwise sailing, where the land distances between point and point were known, they must have almost unconsciously arrived at a scale.

The second ancient pilot-book that has survived, which may have been written at the beginning of the third century of our own era, gives distances, as might be expected, in stades rather than days' sail. This is the so-called *Stadiasmus of the Great Sea*, which is of course much fuller and more precise than its earlier prototype. Like the latter it commences in the Nile mouth, where Alexandria had now become the leading sea-port, and continues westward. 'From Leuce Acte to Zygris 90 stadia: there is an islet: put into the place with it on your left: there is water in the sand. From Zygris to Ladamantia 20 stadia: close by lies a rather large island: put in with this on your right. There is a harbour accessible with any wind: water is to be found.' And in the section on Crete: 'From Casus [Is.] to Samnonium 300 stadia. This is a promontory in Crete which extends far to the north: there is a temple of Athene: it has an anchorage and water.'

In neither of the two ancient books of sailing directions which have survived do we find mention of compass direction (which does not, of course, mean magnetic compass direction), and this is perhaps not surprising, as the visible coast was being followed from point to point. But when open sea crossings were to be made it was necessary to know the direction to take, and from the

IV. This twelfth-century miniature of St Guthlac voyaging to Croyland in the Fens shows a typical medieval ship with square sail, braced mast and starboard steering oar. The pilot has his long sounding-rod for the river journey.

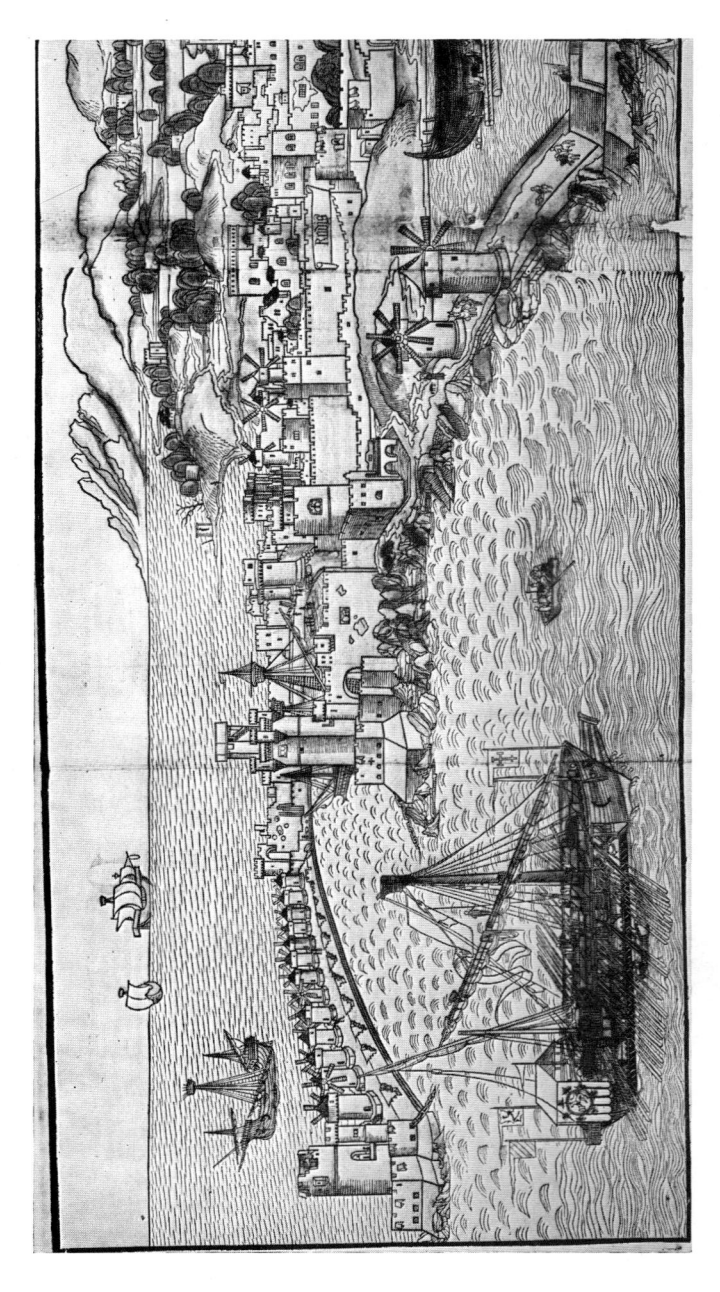

V. Rhodes Harbour was sketched in 1483 by the artist who accompanied Bernard von Breydenbach on a pilgrimage to Jerusalem that year. The galley of the day employed both oars and sails, and lacked the high poop and forecastle of the vessels seen in the distance.

few random references that have survived it appears that this was given in terms of the following wind that would carry the ship to the desired destination. We read, for example, in the geography of Agathemeros: 'From Paphos to Alexandria 3800 stadia with Boreas.' Alexandria in fact lies a little over 300 miles south-west of Paphos, which is on the south-west coast of Cyprus, and Boreas was a north-easterly wind. Again, Strabo who, as already noted, had periploi in front of him, writes: 'From Chios to Lesbos 200 stadia with Notus [the south wind]', and again, 'From Cyrene to Criumelopon [the south-west headland of Crete] 2000 stadia with Leuconotus.' Leuconotus was a wind about 30° west of south, and this direction and distance are approximately correct. Pliny's use of this convention had already been noticed: 'From Carpathus to Rhodes 50 miles with Africus', and we can safely conclude that from the earliest times the mariner's compass, in the sense of his division of the horizon, was the wind-rose, while he learned directions in terms of the necessary following wind. This explains another incident in the voyage which led to St. Paul's shipwreck. The wind that swept the ship away from the south shore of Crete was Euraquilo, a north-east wind veering east. Aquilo was the wind for Cyrene, but with more east in it would carry them past Cyrenaica and into Syrtes Gulf. Here, according to Strabo, were quicksands where no anchor would hold, so that it was to be dreaded, and it was therefore the correct procedure to throw out some sort of drag anchor as they did.

The wind-rose, as already mentioned, might be composed either of twelve winds or only eight, and the former is said to have been devised by Timosthenes, the scholar-sailor who is the next of that type to be known to history after Pytheas. He was a man of Rhodes, chosen by the king of Egypt, the second Ptolemy (285–246 B.C.), to be Chief Pilot of the navy. Practical science and technology (as opposed to the speculative science of Athens) rose to great heights in Alexandria under Greek rulers, and an advance in nautical science is therefore not surprising. Nor can it be wondered at that a Rhodian was chosen as Chief Pilot of the fleet, for Rhodes was then the leader of the nautical world, and is described by Strabo as follows: 'The city of the Rhodians is on the eastern promontory. With regard to harbours, roads,

5—H.A.

walls, and other buildings, it so much surpasses other cities that we know of none equal, much less superior to it. Their political constitution and laws are excellent, and the care admirable with which they administered affairs of state generally, and particularly those relative to their marine. Hence, being for a long period masters of the sea, they put an end to piracy, and became the allies of the Romans. . . .' Among the votive offerings embellishing the city was one of the seven world wonders, the Colossus of the Sun, 70 cubits high, which had been recently erected and was still standing during Timosthenes' lifetime, although it fell after less than sixty years from its erection (Plate V).

The most direct account of Timosthenes' work is to be found in a book written by a late Greek geographer Agathemeros, who after describing the eightfold wind system based on the four cardinals combined with the four solsticial rising and setting points of the Sun goes on to say: 'But Timosthenes, who wrote "periploi" [sailing directions], distinguished twelve winds, putting Boreas between Aparctias and Caecias, Phoenix between Eurus and Notus, Leuconotus between Notus and Libs, and Thracias, which is also called Caecias, between Aparctias and Argestes.' This may be termed the earliest geometrical windrose, with the circle or compass of the horizon equally divided. When the amplitudes of the sunrise and sunset at the solstices are used to fix the positions of the four winds Argestes, Caecias, Libs and Eurus, the four resulting points, in the latitude of Rhodes, are about 30° from the east–west line, or one-third of the quadrant. There can be little doubt that this was the reason for halving the remaining two-thirds, and giving precision to four more vaguely used wind-names. Boreas, for example, had been used, and indeed continued to be used by the laymen, for any cold, blustering wind from due north round towards east. Thracias, which according to Agathemeros was also called Caecias, was now distinguished and transposed to its correct position in relation to Rhodes. Thrace was north and east to the Athenians, and therefore a wind coming from it might be called either Thracias or Caecias; it was on the contrary a country west of north to a Rhodian. The name Leuconotus was generally replaced by Libonotus, that is to say, 'south with Libs' or south-south-west. Phoenix, too, was more usually called Euronotus (south-

south-east), and these appear to be the earliest examples of this simple and logical way of combining two wind directions in order to name an intermediate one, which was to become universal.

Of the sailing directions that Timosthenes wrote the best known was his *On Ports*, in ten books, which his younger contemporary Eratosthenes, who was librarian at Alexandria, embodied almost word for word in his geography, thus laying himself open to a later charge of plagiarism. Others known to have used the Chief Pilot's writings are Hipparchus, Posidonius, Strabo and Pliny, and the twelvefold wind-rose became the accepted system of the classical writers. Nevertheless, there were some who found it, according to Pliny, 'too subtle and meticulous', and were satisfied with two winds in each quarter of the heavens. An eightfold system, in fact, persisted, so it seems likely, among the generality of sailors, whatever may have been the case with the literate, educated minority. There was, too, a very ancient sixteenfold division of the sky and horizon, which was held to be of Etruscan origin, and was used by the augurs, particularly in respect of prognostications about thunderbolts. The eightfold and sixteenfold and later thirty-twofold systems obtained by halving and rehalving the 'airts' are, in fact, much simpler for ordinary men than the wind-rose of Timosthenes, which eventually they superseded. That the Temple of the Winds, still standing at Athens, is an octagon, not a dodecahedron, is therefore significant (Plate III).

The question that must spring to the mind of every sailor is: Did the ancient pilots use a chart? Was the periplous normally accompanied by a map or maps? The only answer that can be given is that it is probable. Certainly the Greeks and Romans used maps freely, and Herodotus describes how an envoy from Ionia showed a route map through Asia Minor and Babylonia to the king of Sparta. And Agathemeros, too, when running over the names of Greek historians and geographers, seems to imply that the provision of maps was a matter of course, not only with descriptions of the world, but with pilot-books and itineraries. A late copy of a Roman road-map, the so-called *Tabula Peutingera*, has survived, and it is clear that a matching sea-chart would have presented no technical difficulty. There are unlikely to have been any parallels or meridians on such a chart, although Eratosthenes was trying to establish them as the framework of his world map, and although it is suggested that Timosthenes was familiar with such lines and himself determined the latitude of Berenice. But it is probable that direction would be indicated merely by putting the named winds, represented by puffing heads, round the margin of the map. Such a gallery of personified winds certainly appears on medieval maps derived from classical sources (Fig. 9), and seems to have been drawn on the map which a poet of the days of Augustus Caesar describes. This poet was Propertius, who imagines a young wife looking at a map or chart as she thinks of her warrior husband away in Parthia. She sees the way the River Araxes flows, and how far a Parthian horse must gallop without water; she sees, too, which lands are hotter, which colder, and finally she sees 'ventus in Italiam qui bene vela ferat'—the wind which will carry his ship safely back to Italy.

Then again, when reading Strabo, who uses the periploi of Timosthenes and others, it is difficult to believe that he has not a map or chart in front of him. In describing the Sporades, for example, he says: 'Carpathus . . . lies opposite Leuce Acte in Africa, which is distant about 1000 stadia from Alexandria, and about 4000 from Carpathus.' And elsewhere: 'Metagonium [in North Africa] is nearly opposite to New Carthage. Timosthenes is mistaken in saying it is opposite Massilia. The passage across from New Carthage to Metagonium is 3000 stades.' Yet again: 'Cossura [Pantellaria] is situated before Cape Lilybaeum [in Sicily], and opposite the Carthaginian city of Aspis . . . it is situated in the midst of the space which lies between those two places, and is distant from each the number of miles last given [88].' And finally: 'In the bay [Laconian Gulf] on the coast is Taenarum [C. Matapan], a promontory projecting into the sea. Upon it, in a grove, is a temple of Poseidon. . . . Thence to the promontory of Phycus [Ras-al-Razat] in Cyrenaica, is a sea passage towards the south of 3000 stadia: and to Pachynus [C. Passaro], the promontory of Sicily, towards the west 4600 or according to some writers 4000 stadia: to [Cape] Malea, towards the east, including the measurement of the Gulf, 670 stadia: . . . to Corycus [Cimarus], a promontory of Crete, the nearest passage by sea is 700 stadia.' Such a description must surely have been read with a chart at hand, even if it had not actually been drawn up from one.

That no Greek or Latin map or chart has survived of any that may have been drawn more than two thousand years ago is hardly to be wondered at. And since after the destruction of the Roman Empire it was Christian monks who preserved what we now possess of classical literature and learning, it cannot be expected that they would have interested themselves in the technical equipment of sailors, or made copies of their sailing directions. Among chance survivals, however, there is another document which is called a *Periplus*, although it is of rather a different character to those already described. It was intended, in fact, as a handbook or guide, not for the shipmaster or pilot but for the merchant-shipowner who customarily sailed as captain of his trading ship. Although the captain was in supreme command, he was not necessarily or usually a professional sailor, and the code of maritime law (traditionally assigned to Rhodes) laid it down that when important decisions involving seamanship had to be made he must consult the ship's officers, and even on occasion take the view of the whole ship's company. It was obviously important therefore that he should have a general knowledge of the sailing conditions, of the wind and weather, the distances and dangers, which governed the voyage he proposed to undertake. He needed to know also something of the native peoples and their political circumstances, as well as the merchandise in demand and the goods available for export at the different ports of call. The *Periplus of the Erythraean Sea* is just such a combination of merchant's handbook and sailing directions as a merchant-captain would require. The anonymous author was a Greco-Egyptian, writing in about A.D. 60, not long, that is to say, before Pliny wrote his famous Natural History, and the area covered is the Red Sea and Arabian Gulf, which is described in considerable detail, with particular attention to the routes to western India, besides some more general notes on the Far East.

The very intimate knowledge shown of Indian waters reflects the volume of traffic that had developed, at first slowly under the Ptolemies, and then much more rapidly when, in the latter part of the first century B.C., Egypt became a Roman province. Strabo

speaks of some six-score ships in the Indian trade in his day and, on the authority of Posidonius, ascribes the first trading voyage to the reign of the second Ptolemy, the king under whom Timosthenes served. The leader was a certain Eudoxus of Cyzicus, a widely travelled man, who took as his guide an Indian trader, the sole survivor of a shipwreck in the Gulf of Aden who had been brought to the Egyptian court. From this time onwards the Greeks and Romans must have had a growing knowledge of the regime of the monsoons in the Arabian Sea. Indeed Eudoxus himself (doubtless employing Arab pilots) was on one of his voyages involuntarily brought back on the north-east monsoon to the East African coast, which had long been frequented by Indian and Arabian traders as far south as the monsoon system extends, that is to say to about Cape Delgado. Eudoxus, however, must have made his outward voyage to the mouth of the Indus by sailing parallel to the Arabian and Makrar shores, for according to the *Periplus of the Erythraean Sea* it was a Greek shipmaster named Hippalus who first sailed across the open sea direct to western India. And the south-west monsoon was therefore called after him the Hippalus. But by the time the *Periplus* was compiled, the three alternative routes, to the Indus, to the Narbuda and to the Malabar coast, were well known.

Pliny describes the first stage of the voyage as from Berenice in Egypt, the passengers setting out a little before or immediately after the rising of the Dog Star, that is to say about the second week in July. Thirty days were allowed for reaching Ocelis, just outside the Strait of Bab el Mandeb, and thence, if 'Hippalus' was well established, they could reach Muziris in Malabar in another forty days. This voyage, it may be noted, required sailing somewhat athwart the general wind stream. The *Periplus* on the other hand describes the voyage as starting from Myos Hormus, not far outside the Gulf of Suez, whence the ship could touch first at Leuce Kome, situated rather farther south on the Arabian shore, whence a caravan route ran to the Roman station of Petra in Transjordan. From that point on, however, Arabia was avoided, and the ship kept well out to sea, in order to avoid the thievish and treacherous tribesmen, until the Burnt Islands (the volcanic Zuqar and Hanish Islands) and civilized Yemen were reached. Here, says the author, there were local merchants, pilots and sailors all familiar with the Indian voyage. To take ship at Myos Hormus had its disadvantages, particularly in respect of the more unfavourable weather, and Strabo relates that his military friend Aulus Gellius lost a number of his ships on this crossing, owing, besides, to the rocks, reefs and tides, for which he must have lacked experienced pilotage.

Interspersed with information about the local peoples, and lengthy detail and lists of the objects of trade, there are paragraphs in the *Periplus* apparently taken from some relevant book of sailing directions. At Moosa, a frankincense port, for example, it is stated that 'the anchorage is safe and good upon a sandy bottom, where the anchors have good holding'; while at the Strait of Bab el Mandeb: 'the current here is violent, and the wind, by being confined between mountains on the two opposite shores, adds greatly to the strength of the current'. This is confirmed by the modern sailing directions for the area which say that the winter monsoon here is south-easterly, and that superimposed on the strong current which the monsoon drives into the Red Sea are powerful alternating tidal currents of up to 4 knots. Among the indications that the ship was approaching the Indus the pilots noted the change in the colour of the water far out to sea, and both here and elsewhere near the Indian coast they saw large numbers of sea-snakes, 'rising up from the bottom and floating on the surface'. The sea-snake, found only in tropical waters, is an air-breather, and often lies stretched out on the surface. Rarely seen in the open ocean, the snakes occur in enormous numbers on the continental shelf, and especially before river mouths, where their food is abundant. Usually it was the commoner smaller species 'coloured bright green, running into gold' that told the pilot the Indian coast was near, but off the Gulf of Baraka (Cutch) the 'serpents' were very large and black belonging to another well-known species of these waters. The dangers of the shallow gulfs of Baraka and Cambay, which lie open to the south-west monsoon, are carefully described. There are shifting sandbanks on which a ship may run aground even before sighting land. Besides tide-rips and overfalls, there are tidal currents so powerful that they may turn the vessel broadside and sweep it forward out of control, while the retreating tide may leave it stranded high and dry. According to the modern sailing directions the normal rise of the tide in the Gulf of Cambay is 25 feet, and it may reach 33 feet at spring tides, while the tidal streams run up to 6 or 7 knots. And even if the Bay before it (says the *Periplus*) is safely crossed, the entrance to the River Narbuda, leading to the city of Barygaza (Broach) on its banks, is only found with difficulty. For the shore is so low as to be invisible, and there are shoals in the mouth of the river. It was for this reason (the writer continues) that the local ruler kept a number of fishermen in his employ who could act as pilots and bring boats up the channel.

This is the first mention of pilotage in local waters, a point to be considered presently. By a strange chance we get a second glimpse of the navigation of the Indian Ocean nearly four centuries after the *Periplus* was written. It comes from the pen of another Greco-Egyptian, a merchant of Alexandria, who in his later years wrote what he called a 'Christian Topography' with the object of proving from the Scriptures that the Earth is not a globe, but is flat and four-square. Cosmas Indicopleustes, as he was called, was familiar with Abyssinia and with Ceylon as a business man, but it was part of his theory that only the great gulfs of the ocean—the Gulf of Aden, the Persian Gulf, the Mediterranean Sea, and so on—were designed by Providence for navigation, and that the ocean itself was an impassable barrier shutting mankind off from the Earthly Paradise. 'Once on a time,' he says, 'when we sailed in these gulfs, bound for Further India, and had almost crossed over to Barbaria, beyond which there is situated Zinguim, as they term the mouth of the Ocean, I saw there to the right of our course a great flight of the birds which they call Souspha, which are like kites but somewhat more than twice their size. The weather was there so unsettled that we were all in alarm, for all the men of experience on board, whether passengers or sailors, all began to say that we were near the ocean, and called out to the pilot, Steer the ship to port, and make for the gulf, or we shall be carried into the ocean and be lost! For the ocean rushing into the Gulf was swelling into billows of portentous size, while the currents from the Gulf were driving the ship into the Ocean, and the outlook was altogether so dismal that we were kept in a state of great alarm. A great flock of the birds called Souspha followed us, flying generally high over our heads, and the presence of these was a sign that we were near the Ocean.'

What the pilot then said and did Cosmas does not relate, and this story drives home a point that partly explains why our knowledge of early navigation is so scanty. Nearly all that has come down to us in writing on the subject was written by passengers and landsmen and not by the seamen themselves. In this case the ship was sailing out of the Gulf of Aden on a voyage to the peninsula of India (not the Further India of today), and was keeping towards the African coast (Barbaria), beyond which to the south, that is to say after rounding Cape Guardafui, lay the region called Zinguim, then the name of the Zanzibar coast. This of course faces the Indian Ocean. The great birds which flew overhead must have been frigate birds and, as these breed in the Seychelles and near Madagascar, it may be accepted that they were one of the signs that sailors noted as an indication that they were approaching East Africa. And if this voyage were being undertaken before the summer monsoon was well established, or in autumn when it had weakened, bad weather, and north-easterly winds combined with the set of the currents, might well carry a ship off its course and round Cape Guardafui. According to the *Periplus* it was usual to keep along the Arabian side of the Gulf, and wait in some harbour near Ras Fartak until the wind was favourable. If for business reasons the ship had taken a less usual course near the African shore the alarm of the passengers at the idea of being carried to Zanzibar instead of India is understandable. Cosmas, however, turned the incident to his pious purpose of Christian topography. He was, in point of fact, an educated man, familiar with such astronomical writers as Ptolemy, and able to discuss the division of the Earth into climatic zones by means of the measurement of shadow lengths. But he was not quite educated enough, and by erroneously postulating a small Sun quite near the Earth was able to support his flat-Earth theory and explain away quite easily the lengthening of noon shadows with higher latitude.

Although it is not until the date of the *Periplus of the Erythraean Sea* that we come across any mention of pilotage waters and local professional pilots, these must always have been necessary once overseas trade was established, especially to bring merchant vessels into estuarine and delta ports. The constant shifting of channels and the building up of mud- and sandbanks is a perennial problem, and such underwater changes are particularly marked in countries like India where there is strong seasonal rainfall, so that the rivers come down in violent flood after nearly drying up. Even in the Mediterranean where the rock-bound harbour was the general rule, there was, for example, the intricate delta of the summer-flooding Nile to be negotiated, and that of the torrential Rhone, swollen now by winter rains and now by the spring melting of Alpine snows. Outside the Pillars of Hercules, too, there were the estuarine rivers, Guadalquivir, Guadiana and Tagus, with tides which could sweep ships far inland, and carry them out again twice a day. A splendid thing, reported Strabo, for the trading vessel, but not without its difficulties and dangers as well. All the same, it was a sore thing for the master of a ship to hand her over to a strange, perhaps a foreign, pilot, and according to ancient maritime law (as written down in the middle ages) if the latter failed in his task and hazarded the ship the sailors had the right to execute him without further ceremony. How far back in time the helpful local fisherman became the professional local pilot it is impossible to say. But for particular voyages skilled men were on hire in Solomon's day, for he obtained Tyrian 'shipmen that had knowledge of the sea' for the fleet he sent to Ophir. When, too, Circe ordered Odysseus to sail to Hades he asked who was going to con the ship on a voyage which no one had ever made. To which the goddess replied that he was not to sit waiting for a pilot but get aboard and sail. She herself supplied the sailing directions. He was to head south ('The north wind will take you along'), and when he had crossed the ocean he would see a low shore, with groves of poplars and willows. These were his landmarks, and there he was to beach his ship. And so the voyage was made.

The modern inshore pilot in his own waters has little need in clear weather of the ship's compass or any other navigating aid. He steers by landmarks and leading-marks, using the fore-and-aft line of the ship, from stem to stern, to get his alignments. In ancient days the pillars and temples that sailors set up on strange shores served them also as landmarks. There was, for example, a very ancient tower to lead ships into the western arm of the Nile, on which stood the then chief city of Sais, and Strabo has the following relation about the mouth of the Rhone. It was a river difficult to navigate 'on account of its great impetuosity, its deposits, and the flatness of the country: so that in foul weather you cannot clearly discern the land even when quite close. On this account the people of Marseilles . . . set up towers as beacons. They also erected a temple to Diana of Ephesus.'

Whether these towers had flares lit on them at night we do not know, but certainly the famous Pharos Tower on Pharos Island near the Nile delta did, and so later gave its name to any lighthouse. The island stood opposite a bay between two promontories, thus enclosing a natural harbour, and the tower was built at the eastern entrance of what was subsequently made into the grand harbour of Alexandria. It was, says Strabo, 'admirably constructed of white marble, with several storeys. Sostratus of Cnidus, a friend of the [Ptolemaic] kings, erected it for the safety of mariners, as the inscription imports. For as the coast on each side is low and without harbours, with reefs and shallows, an elevated and conspicuous mark was required to enable navigators coming in from the open sea to direct their course exactly to the entrance of the harbour.'

That the Phoenician and Greek sailors frequently set their ships' prows freely towards the open sea cannot be doubted or denied. Out of sight of land they steered by the familiar courses of the Sun and of the brightest stars, helped, too, by their knowledge of the 'feel' of the different winds. They knew the approximate distance from port to port, and the way made by their ship. But such navigation was necessarily imprecise, so that landmarks visible at a distance were of the utmost importance. 'My home is under the clear skies of Ithaca,' said Odysseus, 'our landmark is the wooded peak of windswept Neriton.' And in immediate Greek waters, since nearly all the shores and islands are lofty, there are always landmarks visible. A mountain summit of 8000 feet can be seen a hundred miles away on a clear day, and this is the height of Mount Ida in Crete. But there were no such heights in Egypt or in Libya, nor in the long cross-passage of the Arabian Sea. Much, in fact, depended upon those who were making a particular voyage where there was an alternative between coasting —which did not necessarily mean 'hugging the shore'—and taking the direct run. A merchant ship, with skilled officers and a crew of professional sailors, thought nothing of the dangers of the open sea. But when it was a question of transporting an army, or of passengers and even captains unaccustomed to the sea, it was a different matter. Odysseus tells the tale of how he went in a Phoenician ship to Libya. They took the 'central route', running under the lee-side of Crete, but 'when we had put Crete astern, and no other land, nor anything but sky and water was to be seen, the hostile Zeus brought on a storm'. Yet, on the other hand, when the Greeks sailed away from Troy and had reached Lesbos, they held a debate as to whether they should use the sheltered channel inside Chios, which involved, however, rounding the dangerous headland of 'windy Mimas', or take a roundabout outer course from island to island back to Sparta and Ithaca. 'In this dilemma, we prayed for a sign, and heaven made it clear that we should cut straight across the open sea to Euboea.' The heavenly sign was perhaps the whistling north-east wind that sprang up and carried the fleet straight to the southern promontory of Euboea and on to the Sunium promontory of Attica. Even on this direct voyage, however, distant landmarks were not lacking for those with the experience to recognize them. On a much later, and this time actual historical occasion, the question arose of arranging the voyage of the Emperor Caligula (A.D. 12–41) from Italy to Alexandria. In this case the coastwise route was decided on in preference to the direct one, on account not only of the importance of the passenger but of the size and character of the escorting fleet. Fighting ships were not built to carry water and provisions in quantity as were the more roomy merchant ships, and hence they could not risk being several days at sea.

The fact that in the Mediterranean basin, and especially from about Lat. 40° southwards, the summer sailing season is also the dry season, and so the season of clear skies, was an advantage to the early navigator of those regions that cannot be overestimated. Yet the Celtic and Germanic peoples contrived to navigate the very unpropitious seas of north-west Europe, where grey days at midsummer are common enough to excite little remark. Nor did they 'hug the shore', but launched out across the widest part of the English Channel and crossed the North and Norway seas. Indeed they did not hesitate to navigate the ocean itself if need be.

IV

The Irish and the Norsemen

THE Mediterranean Sea and Indian Ocean in which the Greeks and Phoenicians sailed have advantages for seamen which are wanting in the North Atlantic Ocean. In particular the weather is more predictable in the south, the skies are clearer, storms if no less violent are less frequent, at any rate during the summer sailing season: nor are the seas so high. Then, too, the Sun's course above and below the horizon is more uniform as between summer and winter. At Rhodes, for example, little more than an hour is ever added or subtracted at either end of the day with the seasons, whereas in the Shetland Islands the change is nearly four times as great, and round about the summer solstice it is still twilight at midnight. These long hours of daylight are, of course, in one sense an advantage, but the obliquity of the Sun's course, which is their necessary counterpart, means that it is more difficult both to navigate by the Sun, and to use it to tell the time of day. The same is true, besides, of setting course by the stars, for their circles become more and more oblique to the horizon the nearer the pole.

That navigation in the north should lag behind that in the south is therefore understandable, and the little skin-boats that the Carthaginian captain Hamilcar saw in Brittany about 600 B.C. made a strong contrast with his own stout, well-found trading ship. All the same the skin-boats were making (so he reported) the three-hundred-mile crossing to the Holy Island of Ierne, that is to say to Eire. The Oestrymnides (as the native people by Ushant were called) had obviously no dread of the open sea. And three centuries later Pytheas was meeting people who had crossed (we must suppose) from Norway or the Shetlands to Iceland, although what kind of boats they used he does not say. However, by the time that Julius Caesar was master in Gaul there were large oak-built sailing ships in use in the north-west of that country, where the native people were now the Veneti. In the course of the description which he gives of these ships the great Roman general brings out yet another contrast between sailing conditions in Atlantic and in Mediterranean waters, namely the great tides of the former. These came as an unpleasant surprise to southern sailors. As to the ships: 'The keels were somewhat flatter than those of our ships,' wrote Caesar, 'whereby they could more easily encounter the shallows and the ebbing of the tide. The prows were raised very high, and in like manner the sterns were adapted to the force of the waves and storms. The ships were built wholly of oak, and designed to endure any force and violence whatever: the benches, which were made of planks a foot in breadth, were fastened by iron spikes the thickness of a man's thumb: the anchors were secured fast by iron chains instead of cables [warps], and for sails they used skins and thin dressed leather. These either through their want of canvas, or probably because they thought that such storms of the ocean and such violent gales could not be resisted by [canvas] sails, nor ships of great burden be conveniently enough managed by them. . . . Our fleet excelled in speed alone, and in plying of oars.' He goes on to remark that the height of these vessels made it difficult to hurl weapons into them 'and for the same reason they were less readily locked in by rocks. To this was added that whenever a storm began to rage and they ran before the wind, they could both weather the storm more easily and heave-to securely in the shallows, and when left by the tide feared nothing from the rocks and shelves: the risk of all which things was much to be dreaded by our ships.'

The low-built oared galley which Caesar employed for his soldiers was to remain in use in the Mediterranean for another sixteen centuries, nor were his mariners the last to have the unpleasant experience of anchoring in comparatively deep water only to feel the rocks under the keel a few hours later, or even perhaps find themselves left high and dry by the tide. The regime of the tides, in fact, in all its complexity, was something just as necessary and important for the Atlantic pilot to master as was the wind-rose. And nowhere was this more true than in Brittany and round the Channel Islands where there are such terrifying rips and races.

Nothing more is known of the Veneti, but during the long

period of the Roman occupation the people of southern Britain had the opportunity of learning anything that the more civilized invaders could teach them about nautical science. So, too, had the Frisians, the sea-people who lived on the immediate flank of the Roman Empire, between the Rhine delta and the mouth of the Elbe. And their more northerly neighbours, the Jutes, Angles and Saxons, certainly had enough seamanship to make piratical raids on England in large, well-found, oared and sailing ships before A.D. 300. But about all this period very little is known. What is certain is that skin-boats were still being used in the Celtic fringe-lands of the West, and it was in such leather-covered boats that the Irish holy men sailed the ocean. St. Patrick brought Christianity into Ireland in the middle of the fifth century, not very long after the sack of Rome. And as the barbarian invasions were making life more and more difficult for Christian scholars in Gaul and elsewhere, many of them during the next hundred years or so crossed over from the Continent into this quiet, unthreatened outpost of the faith. They brought with them their classical as well as their sacred books, and Ireland bloomed into her golden age of scholarship.

In the Irish monasteries, however, there were many who had a great urge for a life of solitude, and it became a common thing to sail out and establish small religious houses or hermitages on more and more distant islands. In this way some of the monks certainly reached as far as Iceland and in later days legends grew up which centred on the stories of these voyages. Among the most famous and widely popular was the Legend of St. Brendan, which while full of miraculous happenings contains matter-of-fact detail that affords at least a glimpse of the navigating methods used on these extraordinary journeys.

St. Brendan lived nearly a hundred and fifty years after St. Patrick, in a monastery on the Shannon, and the story begins with the visit of a certain Abbot Barinthus. 'Pray, refresh our souls', so Brendan says to him, 'with some of the miraculous things you have seen on the Ocean.' And so the blessed Barinthus relates how, when he was visiting one of his own monks who had sailed away to become an island anchorite, the brother asked him to step into a little boat and look for the Island of the Paradise of Saints. And when they began to row a mist fell, so that they could see neither poop nor prow. But after an hour a bright light shone, and they saw the lovely Island and landed and explored it. But across the middle there flowed a river, and this an angel forbade them to cross. It was this island that St. Brendan in his turn, taking a company of monks, sailed out to seek. Their course is impossible to follow, but they came in turn to an island on which was a monastery, to another on which there was only a hermit and a multitude of sheep, to yet another having only a furlong in circuit, and so on to much more marvellous places—an island of singing birds, a rock on which Judas Iscariot was eternally prisoner, a fiery volcanic island, and an island on which some of them landed and built a fire only to feel it shiver beneath them and to escape just in time from what proved to be a whale's back. Sinbad the Sailor had just such an adventure. Yet again and again there are landscape details that belong plainly to the North Atlantic islands. And the first setting out, in a 'very light boat' which was covered with ox-hide, is most convincingly told. The mast was stepped, and the sail raised, while the provisions they stowed included butter to grease the hides 'as was the custom in those parts', says the story-teller. They set out 'towards the summer solstice', perhaps to the north-east, and as they had a fair wind, that is to say a wind behind them, there was no need to row, or to do anything but hold the sheets. But after twelve days the wind fell, and the monks now rowed until they were exhausted. So St. Brendan bade them ship their oars—even the steering oar (*gubernaculum*)—set the sail, and leave the ship to God, 'who is our pilot and steersman', he said. And towards evening they had a wind: 'But they did not know from which quarter it blew, nor where it was carrying them', says the story. That is to say they lacked the essential pilot's knowledge of the wind-rose—the which, the whence, and the whither, of each of the eight winds. However, this is a miraculous story, and just as their provisions were exhausted the boat reached an island with high rocky cliffs. A narrow harbour revealed itself, and here the travellers were greeted by a dog which led them into a town. In this way they passed from island to island, and whenever they were in any difficulty, as on one occasion when the sea was *quasi coagulatum*, presumably partly frozen, the same order came: 'Ship the oars, loose the sail. Let God steer.' And presently a 'prosperous wind' sprang up, and there was nothing to do but hold the sheets, 'de tenir les cordes' as the French version has it.

While there is occasional mention of the four quarters of the sky, north, south, east and west, the direction of the voyages (there were altogether three) cannot be made out, but after seven years the Island of Saints was discovered, shrouded in mist. It was so large that they explored it for forty days (the numbers seven and forty of course have mystical associations) before coming to the river which must not be crossed. This island lay out in the west, and played its part in the history of navigation, for it was marked, now in this place, now in that, on the maritime charts of later days, and bore the name of St. Brendan's Isle.

All the incidentals of the story—thick mists in which the monks could scarcely see one another in the boat, islands first seen like a cloud resting on the sea, spouting whales, the chorus of bird cries from the island of birds—ring true. And the boat covered with ox-hide was the Irish sea-going curragh, not unlike the Eskimo *umiak*, which has continued in use (although now covered with pitched canvas) down to the present day. It is a matter of historic fact that the Irish of St. Brendan's times were familiar with the western islands of Scotland, where, on Iona, St. Columba founded the first Christian church in Great Britain. And in the life of that saint there is a story of one Cormac who had to run before a southerly gale and was driven north for fourteen days, when he encountered some dreadful stinging creatures who have been plausibly conjectured as the formidable mosquitoes of Greenland. The first firm evidence, however, about far northern travel comes from the writings of a learned Irish monk, Dicuil, who as a young man had taught in one of the schools of the king of the Franks. He wrote a brief *Computus*, or treatise on time reckoning, but his chief work was the book called *De Mensura Orbis Terrarum* based mainly on the standard classical world map and on the writings of Pliny, Solinus, Priscian, Isidore and other authorities. This was written in A.D. 825 and Dicuil tries to correct the very scanty notices of the far north which he finds in these earlier writings. 'There are many more islands in the Ocean north of Britain', he says, 'which can be reached from the northern British islands in a voyage of two days and two nights, sailing direct, with full sails spread, and a favourable wind.' The

6—H.A.

use of this phraseology indicates that the two days and two nights are a measure of distance, not of actual sailing time, that is to say they represent about 200 modern miles, or 240 Roman miles. 'A trustworthy priest told me', Dicuil continues, 'that he had sailed for two summer days and an intervening night in a little boat with two thwarts [i.e. rowing seats] and landed on one of these islands. These islands are for the most part small: nearly all are divided from one another by narrow sounds, and upon them the anchorites who proceeded from our Scotia [i.e. Ireland] lived for about a hundred years. But . . . now by reason of the Northman pirates they are emptied of anchorites, but full of innumerable sheep and numbers of different kinds of sea-birds.' Now the Faroes lie about 200 miles from both the Shetlands and the Orkneys, and could have been reached from either by Dicuil's informant under favourable conditions in the time that he mentions, and there is every reason to believe that this is the archipelago described (Fig. 10). And as there is no suggestion that the priest told his tale at the very time Dicuil was writing his book, nor that he had visited the islands immediately after the anchorites had deserted them, the recorded century of occupation goes back to something like A.D. 700. Nor would it have been any less easy for St. Brendan to have visited the Faroes a century earlier still or less likely for him to have found anchorites there. But Dicuil had an even more interesting report to make about the north. He had met certain trustworthy clerics who told him that thirty years before he was writing, that is to say in about A.D. 795, they had spent a spring and summer in Thule, from February until August. There, as they told him, it was light enough at midnight on the day of the summer solstice, and even for a few days before and after it, for a man to do any sort of work he chose, and even to pick the lice out of his shirt!

A long stay in Iceland such as this must surely have had for its purpose the visiting of a Christian community already there. And in fact when the Norse Víkings first reached Iceland in A.D. 870 they found 'Papar' there, as they termed monks and priests, and other men who had come over the seas from the west, that is to say from Ireland. These people left behind them 'bells, books and croziers', says the writer of the *Landnamabók*, and there is confirmatory evidence of the facts from the place-names which

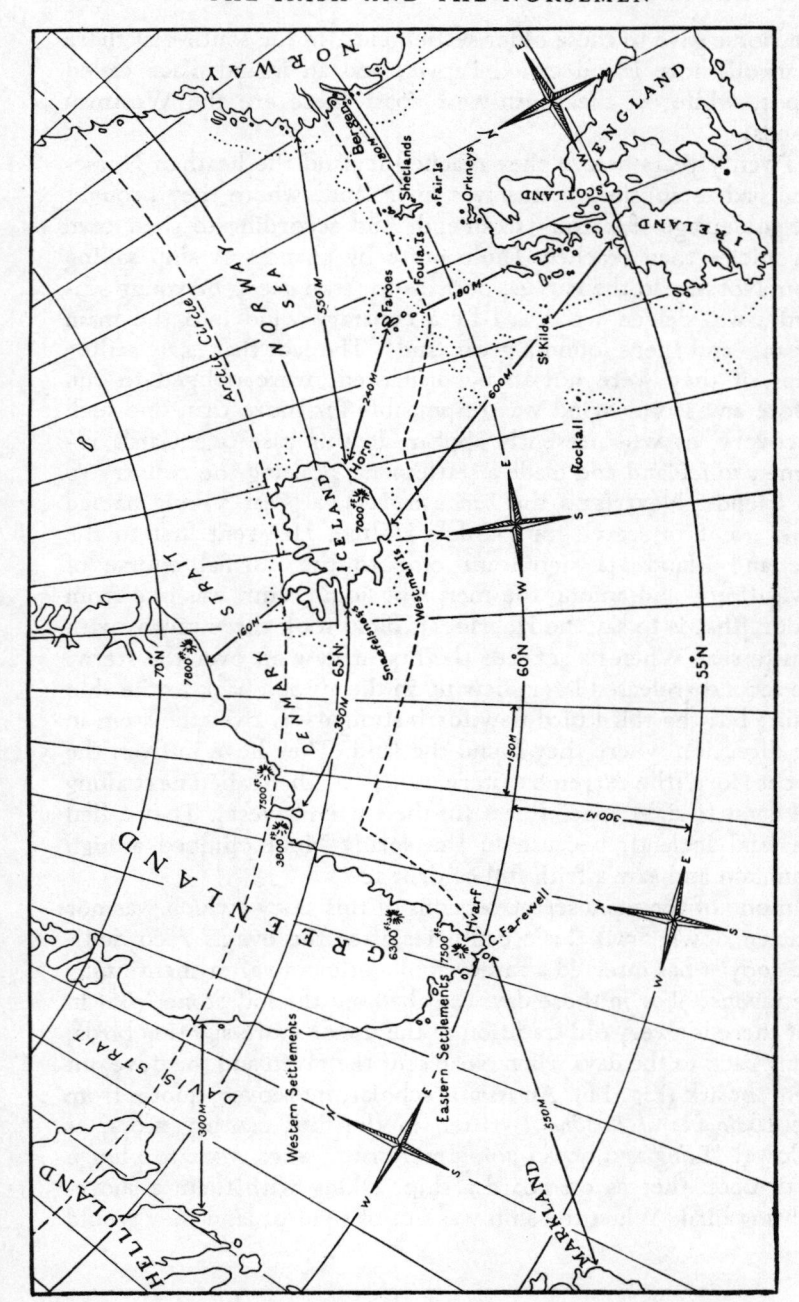

Fig. 10. Early Norse and Irish voyages. The 100-fathom line is shown dotted off N.W. Europe.

the Norse gave to these older settlements. In the south-east there is an off-shore island called Papyle, and an inland place called Papos, while off the south-west coast there are the Westman Islands.

Twenty years before they reached Iceland the heathen Norsemen had established themselves in Ireland, where they brought the golden age of learning to an end. And according to their own chronicles they reached Thule quite by chance. A ship sailing from Norway to the Faroes, perhaps to fetch sheep or young seabirds, was driven westward by a contrary wind into the main ocean, 'and there found a great land'. The fact that early sailing ships, if they were not to be dismasted, were obliged to run before any strong wind was responsible for more than one such discovery, as will presently appear. In this case one Garda returned to Iceland and made a settlement, praising the country to his friends. Next (says the *Landnamabók*) 'a great Viking named Floki went to search for Garda's Holme. He went first to the Shetland Islands [a significant clue to the normal course of navigation], and among the men who joined him was one from Sodor [that is to say the Hebrides]. Floki took three ravens with him to sea. When he set free the first it flew aft over the stern: the second (released later) flew up in the air and back to the ship again; but the third bird flew forth straightway over the stem in the direction where they found the land. They hove in from the east at Horn [the extreme eastern cape] . . . then sailed next along the coast to sight Snaefelsness [in the extreme west]. They called the land Iceland, because in the spring Floki climbed a high mountain and saw a frith full of drift ice.'

In one of the manuscript versions of this story (which was not written down until three centuries after the events recorded), the copyist has inserted a rather apologetic note after mentioning the ravens: 'For in those days they had not the lodestone.' But in fact there is a very old tradition of the use of shore-sighting birds, going back to the days when Noah sent the raven and the dove out from the ark (Fig. 11). An Asiatic scholar, moreover, quotes from the *Dialogues of Buddha* (written in the fifth century B.C.), as follows: 'Long ago ocean-going merchants were wont to plunge forth upon the sea on board a ship, taking with them a shore-sighting bird. When the ship was out of sight of land they would

Fig. 11. Noah makes use of shore-sighting birds.

set the bird free. And it would go to the east, to the south, to the west and to the north (circling, that is to say, with the Sun) and to the intermediate points, and rise aloft. If on the horizon it caught sight of land, thither it would go, but if not, it would come back to the ship again.' This story is exactly confirmed by Pliny's account of the sailors of Ceylon. 'They take no observations of the stars in navigating,' he says, 'the Great Bear is not visible. But they carry birds with them, frequently liberating them, and following the course they take to reach the land.'

Floki's voyage to the Shetland Islands could be made in two days' sailing with an easterly wind from Bergen, from which it lies about 200 miles west. Then from the Shetlands to the Faroes is another 200 miles, or rather more, and this second island group is mountainous, and would be sighted forty or fifty miles away on a clear day. The following wind needed would be from the southeast. The third stretch of the voyage, from the Faroes to Iceland, is a good 250 miles, requiring the same following south-east wind, and the mariners would now have to sail for at least a day out of sight of land. It was here then (if the story is true) that they would need the raven to guide them. Actually, if they were steering on any bearing between 30° and 80° west of north, they would find themselves in soundings on the third day, and in sight of the icefields and mountains of Iceland. On the other hand if the wind had veered and sprung up strongly from the east and northeast they might have been swept right past Iceland, as Gunbjorn was to be about a century later. This Viking was in fact able to get back to port again, but not before he had sighted the islands off Greenland which became known as Gunbjorn's Skerries. It was on an impulse to rediscover these Skerries that Eric the Red set out in A.D. 985 from Snaefelsness, the lofty peninsula south of Breidfjord, and very nearly the most westerly point of Iceland. Following the westerly trend of this peninsula he reached the Greenland coast, says the old saga, 'near Midjökul, which is now called Bláserk: thence he sailed south along the coast to ascertain if it was habitable there'. Bláserk, in most versions of the story, is called Hvitserk, and under this name became the recognized sailing-mark for Greenland. It was probably in the neighbourhood of Angmagssalik, but it has not been identified, and 'before discussing the old Norse sailing-directions that have survived it is worth returning for a moment to the Irish monks. No accident of being driven westwards could have brought them involuntarily, as it brought the Vikings, to the Faroes, Iceland or Greenland.

From Malin Head in north Ireland the distance to Iceland is 600 miles in a north-westerly direction. Sailing from the Head on any bearing between 15° and 45° west of north would bring a ship in sight of the mountains of Iceland (supposing the air was clear and it was daylight), and into soundings, within six days—that is to say, of course, if the sailors had a following wind. And the steersman would know that if he came near to Rockall he must be too far to the west, or if near to St. Kilda he would be too much to the east. Alternatively, he might sail due north from the Head, skirting the Outer Hebrides until the Butt of Lewis was reached. From this point, with a favourable wind, he could lay any course between 20° and 60° west of north, and after covering 400 miles find himself in soundings somewhere off the Iceland coast. And in south-east Iceland, not far from the sea, Ovaeforjokul rises to nearly 7000 feet, so that even from a little curragh it could be seen on a clear day perhaps ninety or even a hundred miles away. The winds, between southerly and easterly, which would be needed for the voyage are usually associated with fine weather during the summer months, and if they were blowing out from an anticyclone (or high-pressure system), they might remain steady for several days together. The north-westerly winds, serving for the return journey to Ireland, occur in the rear of every passing depression, and although they are gusty, or even violent, there are clear, bright skies between the cumulus clouds and thunderstorm clouds that accompany them. The voyage to and fro, although dangerous and adventurous, was not foolhardy. When a 'depression' crosses the British Isles the south-westerly and westerly winds here are likely to have as their counterpart easterly and north-easterly winds across the Norway Sea. These could be counted on with the greatest certainty for the voyage to Iceland early in the sailing season, during the latter part of April and in May. And between Iceland and Greenland they are the most frequent winds all through the season, although along the Greenland coasts themselves the winds are more variable. And it is one advantage that the northern sailor has over his Mediterranean rival that there is no regular dry season, so that even if he is driven off course he can catch rainwater in a weighted sheet of canvas when his supply runs out. But the storm hazards are great.

Even if it can be shown, however, that the voyage to Iceland was feasible the question arises: How did the monks know that such an island existed? They could read, of course, the scanty references to Thule in the Latin works of Pliny, Solinus, and Martianus Capella, but such a schematic world-map as was drawn by the last of these three (of which copies have survived) could never serve as a sailing chart. It has been very plausibly suggested, however, that their guide was the annual migration of geese between Ireland (as well as other parts of Britain) and Iceland. These noisy and conspicuous birds, the brent-geese, white-geese, barnacle-geese and grey-lags, wintered in their tens of thousands in Northern Ireland, and in the Shannon Estuary (which St. Brendan knew). They fly into their feeding grounds, coming in from the north, towards the end of September and during the month of October. But ten days or a fortnight after the spring equinox they begin to leave. A large flock rises, and one long skein after another disappears over the northern horizon. A week or so later another flock departs in the same way, honking as they fly, and in four or five weeks' time the marshes are deserted—all have disappeared. They are on their way to summer breeding places in Iceland and Greenland. Flying along a direct route at about 30 miles an hour they will reach Iceland within twenty-four hours (Fig. 10). Those which continue on the same line of flight across Iceland and then across Denmark Strait can rest if they choose before this second sea-crossing. But it is barely two hundred miles, for at Angmagssalik the Greenland coast bends north-eastwards. Whatever fantastic stories about the birth of barnacle-geese from trees the stay-at-home writer might repeat, the actual sight of these goings and comings year after year was to sailors convincing evidence of a land lying to the north. And the greatest authority on geese, Commander Peter Scott, states that there is nothing impossible in the monks' setting course by the northward migration, and keeping it by watching successive skeins of birds pass overhead. If the wind failed they could ply their oars, as St. Brendan's companions did, and in a week or less Iceland would lie on the horizon. The migrations, besides, both to and fro, coincide with the beginning and ending of the sailing season.

No land-sighting raven would serve on a voyage of 600 miles, but there is ample evidence of the close attention that sailors always paid to the flight of birds. The story of Cosmas in the Gulf of Aden has already been told, and the Portuguese historian Azurara relates that his countrymen discovered Flores, lying 140 miles beyond Fayal in the Azores, because they noticed flocks of land-birds flying in that direction and followed them. A modern writer remarks that the sight of large numbers of eider-ducks is a sure sign of land within a hundred miles, and the nightly home-coming rush of fulmars is but one of the aids that a sailor has under an overcast sky to tell him where the land lies.

In the records of the old Norse voyages, however, there are only two references to the use of birds, one of them a mere hint. The more important of the two is contained in the sailing directions which were given for the very remarkable direct crossing which used to be made all the way from Norway to Greenland, a distance measuring over 1500 miles across the ocean. The starting point was Hernoe, near Bergen, almost the most westerly point on the Norwegian coast. Sailing out westwards, the pilot was told to keep so far north of the Shetlands that they were only just in sight. He is then to sail south of the Faroes, at a distance at which the heights are half-hidden by the sea. The most southerly of this group, Siderö, has lofty cliffs up to 2000 feet, and mountains not far behind. The track may be estimated at about forty miles to the south, and would lie roughly along Lat. 61°. But in the next stage the ship is to keep south of Iceland 'so that the sea-birds and whales can be seen', at a distance which has been variously interpreted, but must be only sixty miles or less. This would mean that the intended track was to lie above the steep submarine slope which borders the 'continental shelf' or pediment from which Iceland rises. Here, according to zoologists, there is an extraordinary proliferation of marine life, with myriads of the tiny organisms which make up the 'crill' on which the 'right' or Greenland whales feed. It would have been therefore a gathering ground of these conspicuous whales (fishing has destroyed them today), besides innumerable fish, and preying upon these a riot of sea-birds. Nor would this be all, for the abundance of herrings and squids, which also feed on crill, make the same belt a favourite feeding and spawning ground of the bottle-nose whale. It would be easy to keep along the oceanic edge of this unusually lively belt of the sea, that is to say along about Lat. 62°. Once the surrounding shelf of Iceland was passed, the crossing was again over the open ocean. The helmsman was to continue to steer west across the now empty, lifeless, cold waters of Denmark Strait until Hvitserk was sighted, and then turn to the south-west (or *utsundr*) and follow that direction until Hvitserk lay due north. At this point the ship could be steered in to the Greenland coast, making for the high land of Hvarf, which was the Cape Farewell region behind which there are heights up to 7000 feet. The Cape was then rounded and the settlements, which lay on the south-west coast, were reached. The reason for not sailing directly in towards Hvitserk (as Eric the Red had done) was that a tongue of ice extends south until late in the season on the far side of Denmark Strait, and this ice barrier seems to have become more extensive during the medieval period (Fig. 8b).

The second mention of birds occurs in a story related about the early days of the settlement of Greenland. According to the saga, Thorstein Ericsson set out from his father's home to follow up the discovery of Vineland made by his brother Lief. 'They were long tossed about upon the ocean: they could not lay the course they wished. They came in sight of Iceland, and likewise had birds from Ireland. Their ship was, in sooth, drawn hither and thither over the sea. In the autumn they turned back, worn out.' The implication is that they had westerly winds instead of the necessary north-easterlies, so that the ship was driven at one time north-east to a point within sight of Iceland and again south-east until the sea-birds flying out from the west coast of Ireland were seen. This might be anything between fifty and a hundred miles from the Irish coast. 'Even when you are a long way out of sight of land,' says another modern writer who has sailed in northern waters, 'the majestic flight of the homing gannets will show you where the land lies.'

The old Norse sailing directions were first collected by Archbishop Erik Valkendorf in the early sixteenth century, long after the regular sailings to Greenland had been abandoned. And one of them contains the puzzling statement that when sailing from

Iceland to Greenland on a clear day both the high Fjeld on Snaefels Jokul and the landmark Hvitserk could be seen at once when the ship was halfway across. This is not considered possible during the 350-mile crossing due east, unless indeed abnormal refraction (not uncommon in high latitudes) raised the Greenland peaks up above the horizon. But on the 160-mile crossing between the north-west extremity or Horn of Iceland and the opposite Greenland shore, where there are some unusually lofty peaks, modern sailors say that on occasion both lands have been visible to them at one and the same time. Another ambiguity in the sailing directions is introduced by the double meaning, still current of course in our own language, of the word 'day'. This may be a period of twenty-four hours when we use it in a calendar sense, but each calendar day contains a 'day' between sunrise and sunset, and a night. Or, alternatively (as Chaucer reminds us), a 'day' for peasants lasts, and is reckoned, from dawn to dusk. The 'day's sail' of a Norse saga or rutter may need to be reckoned at two 'days' in twenty-four hours, since the ships did not tie up at night, or it may mean the conventional 'day's sail with a following wind' which was a measure of distance and not of actual time. The crossing from Stad in Norway to Horn in eastern Iceland is given as seven days' sail, and this suggests merely an estimated distance of about 700 miles, the actual distance being rather over 600 miles. The Norse voyages to the Faroes, Iceland and Greenland were made in round-built merchant ships, and not in the long-oared Viking ships, used for piratical raids and war-like purposes, of which alone specimens have been preserved.

That both the Irish and the Norse voyagers could orient themselves by the noon Sun and the Great Bear as Odysseus' sailors did must be accepted, although only one mention of this procedure is to be found in the sagas. The occasion was during a voyage made by one Bjarni Herjulfsson, who was by profession a trader and spent alternate winters with his ship away in foreign lands. Returning to Iceland one summer he found that his father had sailed that spring with Eric the Red to the new Greenland settlements. Bjarni decided not to unload his ship but to follow his father. He first consulted his ship's company, and all agreed to go, although as their master said: 'Our voyage must be regarded as foolhardy, since no one of us has ever been in the Greenland seas.' However, they set sail and after three days lost sight of Iceland which was 'hidden by the water', but then the 'fair wind' died down. Northerly winds arose, bringing fog. The crew handed all sail, and they drifted about for several days without knowing where they were going. Then the sky cleared, they saw the Sun again, and so *ok máttu þa deila aettir*, that is to say they were able to take their bearings, and find the airts or divisions of the horizon. These were eight in number as has already been described, and the corresponding eight wind-names are the only ones used in the sagas. Once they had recovered their direction Bjarni and his company hoisted sail and set course west again, so that they accidentally made the first discovery of the coast of the American continent, for they had got too far south for Greenland. And they would realize this as likely (for they coasted the new continent northwards) when they found that the nights were distinctly longer than in their homeland Iceland. Bjarni as a trader would be familiar with the relation of the changing length of the day to latitude (although that word would not be used) for, as a young man aspiring to the same profession put it in a book of later date: 'Seafaring traders ought to note such differences precisely, so as to determine what seas they are upon, whether they lie to the north or to the south.' And in fact when Lief Ericsson followed up Bjarni's report of the strange coastline and actually wintered in Vineland, the saga states that the days and nights were more equal in length there than they were in Greenland, so that on the shortest day both *dagmalstad* and *eykstad* lay within the daylight hours. The controversy as to the precise latitude that this implied need not concern us here, but only the observation.

That by the ninth century at latest northern sailors were already using our Pole Star to give (Fig. 5) them a more precise north than merely looking at the Great Bear may be inferred from the Anglo-Saxon vocabulary, which contained the words *scip-steorra* or ship-star, and *lád-steorra*, that is to say lodestar or leading star. It is true that Polaris, the extreme tip of the tail of the Lesser Bear, was then about 7° from the true celestial pole of rotation, but there was no question of actual measurement, and to the northern observer this star stands alone in that part of the sky, while as the Great Bear wheels round, the two pointers always show its position, and distinguish it from Kochab, the only bright star at all near to it. Kochab, moreover, has his less brilliant but clearly visible 'brother', whose changing position, now beside, now above, and now below him shows that he moves, and is not changeless and reliable as the lodestar seems to be. The astronomers, of course, knew better than this, and in A.D. 996 the great mathematician Gerbert, afterwards Pope Sylvester II, wrote to a friend about using a fixed tube to determine whether a particular star was motionless or not. If it were not precisely at the celestial pole it would in a short time move out of the field of vision. Gerbert, however, stands on the threshold of the first revival of learning, during which navigation underwent a major change through the introduction of the magnetic compass. In the period of sailing under review there is no direct evidence of the use of the North Star, although it may be inferred from the simple fact that ships sailed freely through the night.

A lively picture is drawn of what the skilled seaman was expected to know in an old Norse handbook called *The King's Mirror* (*Konungs Skuggsjá*), which was written by an unknown author in about A.D. 1250. While the date is late, the fact that no mention is made of the lodestone makes it reasonable to use the book in connection with the primitive period when the seaman had no magnetic compass. It is written in the favourite dialogue form, an elderly father giving advice to his son who wishes to become a sea-trader. As merchant-owner he will take the rank of captain of his ship, and although he will not have to navigate her himself he will have to know a good deal about seafaring in order to make the right decisions. This son is the young man, in fact, who agreed that seafaring traders ought to know precisely the conclusions to be drawn from the length of daylight in a particular place. In addition, his father tells him, he must master arithmetic, observe the courses of the heavenly bodies, recognize the quarters of the horizon, mark the movements of the ocean and understand the rise and fall of the tides.

The elder man's astronomy is somewhat vague, but he appears to have read a *Computus*, a work on the reckoning of time and the making of a calendar, such as the Venerable Bede had compiled, largely from classical authors, in about A.D. 725. Bede's Latin work was translated into Anglo-Saxon, and was the basis of many similar works of later date, including the Norse *Rimbegla*. It contained a clear account of the monthly cycle of the tides, and on this subject the *Mirror*, too, is specific: 'The tide when it rises completes its course in seven days and half an hour of the eighth day. And every seven days there is flood tide in place of ebb.' 'Half an hour' should read 'half a day', for the daily retardation of the tides brings flood in the place of ebb when it has totalled 6 hours or 360 minutes. If this is divided by 7·5, it gives a daily retardation of 48 minutes, which was the accepted figure. The account continues: 'The Moon when it waxes completes its course in 15 days less 6 hours, and in like period it wanes until the course is complete and another comes. And it is always true that at this time [that is to say every $14\frac{3}{4}$ days] the flood tide is highest and the ebb strongest.' This means that the spring tides occur at this interval. 'But when the Moon has waxed to half, the flood tide is lowest and the ebb, too, is quite small.' That is to say, neap tides occur. The relation between the age of the Moon and the tide cycle must have been known empirically to the early sailors and pilots for their native ports, but they could not, of course, formulate it as the scholars did, and the use of tidal diagrams and tables belongs to a later age and chapter of this book.

Perhaps the most illuminating pages of *The King's Mirror* are those in which the father explains the divisions of the horizon to his son, for he identifies them with the eight winds, that is to say with the wind-rose. And all the winds are personified, suggesting that he had in mind a world map in which each is drawn in the margin as a differing human head (Fig. 9). His rhetorical and picturesque style makes the passages far too long to quote verbatim, and here they will be summarized. The son opens the subject by asking why the sea, sometimes so 'blithe', turns wrathful and dangerous, for the course of the Sun does not appear to explain it. And the father replies: 'The Sun at first visits the east with warmth and bright beams, the day lifts up a pleasant face to the east wind, who turns a bright countenance to his neighbours. In succession the Sun visits the south-east wind, the south, southwest, west and north-west winds, so that all are bright and warm. But the last named then sends a shadow over the Earth to proclaim the hours of rest. At midnight the north wind goes forth to meet the coursing Sun and leads him through rocky deserts towards sparse built shores. He calls forth heavy shadows, and with cool lips gently blows upon the face of the south wind. On the coming of morn the north-east wind leads forth the gleaming day [the description is of the summer season] and soon the Sun rises. Then the sea begins to bar out all violent storms and make smooth highways. There is a covenant among the eight winds which calls forth all the delights of Earth and sky and the calm stirring of the sea. And while this covenant holds', adds the father, 'there will be fair sailing for you.' But about the middle of October there is a change, so that one should not any longer venture out into wide seas, or into those made dangerous by currents, breakers, rocks or sand-bars. 'For the east wind begins to think himself disgraced now his golden crown is taken away [the Sun is now rising south of east], he puts on a cloud-covered hat and breathes heavily and violently. The south-east wind, seeing his neighbour's grief, knits his brows under clouds and blows froth violently about him.' Then the south wind puts on a cloud-lined mantle and blows vigorously, while the south-west wind 'sobs forth his soul's grief in heavy showers, rolls his eyes above his tear-moistened beard, puffs his cheeks under his cloudy helmet, blows the chilling scud violently forward, leads forth huge billows, wide-breasted waves and breakers that yearn for ships, and orders all the tempests to dash forward in angry contest'. Next, 'the west wind, deeply grieved, pulls on a cloud-grey cloak and breathes heavily while the ill-tempered north-west wind, who has suffered the loss of his evening beauty [the summer sunset], knits his brows fiercely, throws rattling hail violently about, and sends forth rolling thunder, with terrifying gleams of lightning. Now the north wind brings out a dim sheen glittering with frost, places an ice-cold helmet on his head, above his frozen beard, and blows hard against the hail-bearing clouds. The chill north-east wind sits wrathful with snowy beard, breathes coldly through his wind-swollen nostrils, and blows piercing drift-snow vigorously forth.' It is clear that except for short journeys of a day or two sailors should stay at home, but, says the father, men may venture out again upon almost any sea except the largest as early as the beginning of April. For since mid-March the days have lengthened, and the Sun rises higher, while peace is now renewed among the winds and the waves sink to rest.

The unknown writer of the *Mirror* shows a remarkable knowledge, not only of wind, weather and tides, but of the movements of whales and fish, of walrus and seals, and of the character and movements of sea-ice. The seals, he says, follow the ice 'as if abundant food would never be wanting there', as indeed is the case. The melting edge of the sea-ice, in fact, swarms with seal and walrus, the latter yielding the ivory and the tough hides for ships' cordage which were once the most valuable products of the Arctic. The monk Dicuil had something to say about the ice edge. He was indignant about such nonsensical statements in the textbooks as that Thule had a six-months' day and a six-months' night, and was surrounded by a frozen sea. The priests whose visits he described had been there (he says) as early as February and found a normal alternation of day and night, while they also found that the ice edge was a day's journey from the northern coast. This is its average position today, but occasionally it closes in on the northern and eastern shores of the island, and it may have been on one such occasion that Pytheas saw its margins and spoke of the 'sea-lung'.

It has been mentioned in passing that all the written accounts of the Irish and Norse voyages (except that by Dicuil) must be dated long after the events which they relate, which naturally adds elements of doubt to their correct interpretation. There was, however, one Viking voyage which was put into writing almost immediately, at the dictation of the man who made it. This was the voyage from Halgoland in Norway round the North Cape to the White Sea which was described by Octhere to King Alfred, and added by the latter to his translation of Orosius' History into Anglo-Saxon. The date of the voyage was about A.D. 880, that is to say soon after the first settlement of Iceland. Halgoland lay behind the Lofoten Islands, and Octhere's farm or estate was in about Lat. 68°, 'the farthest north of any Northman', he said. His description of the course he took is in the simplest possible terms—so many days' sail north, so many east, and so many south. Throughout the voyage he had the land on his starboard, that is to say on the right side of the ship, to which the steering-oar was fastened, while the ocean lay on his 'backboard' or larboard. Consequently it was only necessary to specify the direction followed in very general terms. He said he followed a 'north-right' direction for six days, which was three days' sail beyond the normal limit of the walrus-hunters' voyages. Then the direction of the coast changed, so that he was probably at Nordkyn. Actually his previous direction had been practically north-east, and while he said that the coast now turned 'east-right' the fact is that it runs east-south-east between Nordkyn and Swjatoi Noss. This second section, too, which he measured as four days' sail is only about 350 miles in a direct line and not 400 miles. The first six days' sail involved rounding many capes and islands and so is not easy to measure. For the third section, right into the White Sea, Octhere said he had to wait for a 'north-right' wind, although actually the coast curves now rather east of south, and now south-west. But it is quite possible that the terms *utnordr* and *landnordr* would have been strange to Alfred and these generalizations were reasonable. The point of great interest however is that when he was at Nordkyn the Viking had to wait for a wind which he said was 'westan odde hwôn nordan', that is to say 'west and somewhat north', which shows that he was aware that the true coast trend was rather south of east, and illustrates also the natural turn of expression that later was to crystallize out into 'west by north'.

Such a crystallization only became possible when the magnetic needle allowed of a precise division of the horizon into 'points' rather than arcs, as will appear in the following chapter. Meanwhile a description of the *Mu'allim* or Pilot in the Arabian Sea, originally written in Sanskrit in A.D. 434, sums up the equipment of the skilled seaman of what may be termed the first or primitive navigation period. 'He knows the course of the stars and can always orient himself; he knows the value of signs, both regular, accidental and abnormal, of good and bad weather; he distinguishes the regions of the ocean by the fish, the colour of the water, the nature of the bottom, the birds, the mountains, and other indications.' And the sole implement of his craft was the sounding line, with perhaps, if he were Greek, a crude chart or a periplus.

PART THREE

WITH COMPASS AND CHART

V

In the Mediterranean Sea

THE year A.D. 1000 is a convenient date to choose to mark the beginning of an upward trend in European civilization following some five centuries of barbarism and political confusion. These centuries are often termed the Dark Ages, and they are certainly dark in respect of the history of navigation. Nothing can be gleaned as to the techniques of Mediterranean sailing after the destruction of the Roman Empire, and if these did not deteriorate they could only have stood still. Yet it was during this very period that the Irishmen and the Vikings in their wild northern waters showed what striking feats of seamanship could be accomplished with no other helps than keen observation and stored experience of the nature and behaviour of sea, wind and stars. Advances beyond such simple skills were made possible by two events which occurred about A.D. 1000 or not long afterwards. The first was the revival of the study of mathematics and astronomy in the west, and the second was the discovery of the directive property—the so-called polarity—of the magnet stone or Heraclean stone of the ancients, actually a lump of a quite common sort of iron ore.

It is not surprising that when education first fell almost entirely into the hands of the clergy, at monastic and cathedral schools, its scope was narrowly restricted to the preparation of boys for office either in the Church or in administration and stewardship. All the mathematics they needed was sufficient arithmetic to keep household and estate accounts, and sufficient astronomy for the computation of the calendar. And in western Europe, where Latin was the language of scholarship, the great Greek mathematicians were unknown, or known only at second hand through the inferior writings of late Latin authors. Even in the Greek Byzantine Empire, where a man like Cosmas Indicopleustes could cite familiarly the great Alexandrine astronomer Ptolemy, the pursuit of science and mathematics was left largely to the Syrians.

Great churchmen decried the old pagan learning, but the Syrians had become Nestorian heretics, and perhaps this (and their ancient Phoenician tradition) accounts for the fact that they translated the more important Greek texts into Syriac, and that we owe to one of them the oldest treatise on the astrolabe. And, as it happened, Syria was the first country to be conquered by the Arabs, a people who found mathematics and astronomy very much to their taste. Consequently they re-translated the Greeks from Syriac or translated them afresh, and because they had not the Greek prejudice about manual work they developed the art of making and using mathematical instruments to great perfection. Observatories were set up and libraries collected under royal patronage, while many Jewish scholars living among the Arabs played an important part in this intellectual renaissance.

By A.D. 1000 the main period of Muslim advance and conquest was over, and the Europeans had recovered much of the territory in their own continent which the Arabs had gained. Western rulers thus found themselves with Arab subjects, and, perhaps even more important, with Jewish subjects, who could become their teachers. Christian scholars visited Sicily and Spain, and the great period of re-translation of such writers as Euclid and Aristotle from Arabic or Hebrew into Latin was begun. It became possible for a select few (for most people detest mathematics) to love and master geometry and astronomy, even though to do so excited suspicion. As Roger Bacon remarked, many religious people put mathematics among the Black Arts, and even Pope Sylvester II, the former Gerbert, was dubbed a magician because of his star-gazing and his new Sun-clock. We read, however, that one Ragimbold, described as 'magister' of schools in Cologne, wrote about 1024 to his friend Radolf, who taught in the cathedral school at Liége, to tell him that he had acquired an astrolabe, and to ask him to come over and see it. It was much too precious to lend, for a medieval astrolabe was an elaborate and costly instrument, beautifully made, which served for time-keeping, as well as for all observations and calculations necessary for astronomy, astrology and calendar-making.

And this new-found delight in mathematical practice even reached as far as Iceland, to judge from the recorded observations of a man whom his fellow-countrymen proudly called Star-Oddi.

Oddi lived on the north coast of Iceland, and had no astrolabe. He did not even know the division of a circle into degrees, but he made a surprisingly accurate table of the Sun's noon altitude at fortnightly intervals, using its own diameter and semi-diameter as his measuring scale. Moreover he made a primitive table of amplitudes, that is to say he recorded the day of the year on which the first glimmer of dawn and the last glimmer of dusk was to be seen at each of the eight wind-points on the horizon, and at each of the mid-points between them. On December 30th, for example, he says that 'the day came up' midway between east and land-south (i.e. east-south-east) and went down midway between west and out-west. Forty-three nights later dawn appeared in the east, and dusk disappeared in the west. Twenty-five days later day came up between east and land-north, and so on through the year. This of course was to give mathematical precision to the facts of general experience about the Sun's course which ordinary sailors and fishermen memorized and made use of for their rough reckoning of time and direction. There is no evidence that Oddi's observations, made for approximately Lat. 66° N., were actually used by seafarers. Nor indeed were astrolabes used except by landsmen; and Pope Sylvester's sound conclusion that the Pole Star did not stand at the celestial pole made no impression on the general public, nor did it reach the sailor's ears at all. Nevertheless all these happenings indicate a gradual change in the climate of thought between A.D. 1000 and, say, 1200, and it was this change which eventually made it possible to introduce mariners to instruments and to observations of precision.

It is important to remember that the Eastern or Byzantine Empire continued to prosper and its ships and sailors to trade and fight in the Aegean and Black Seas right down to the days when Anatolia was conquered by the Turks in 1071. And even after that date (since the Turks were landsmen and kept to the plateau), the huge city of Constantinople continued to hold its unique position as a mart between east and west in which merchants and seamen of all nations from t' ree continents met and trafficked. Centuries earlier the Emperor Justinian had reconquered Italy from the Goths and made it part of his Empire. But subsequently the barbarous Lombard invasion of A.D. 568 had driven Italian refugees into the marshes of Venice where their only resource was the sea. Such seaboard cities, too, as Amalfi, Pisa and Bari were then cut off from the interior, and so driven to concentrate on seafaring for a livelihood. Still vassals of the Byzantine Empire, and enjoying protection and privileges in consequence, these maritime communities became in effect independent city states and mutual rivals. Amalfi at first stood in the lead, but by 1154 she had been destroyed by crushing blows from Pisa. A generation later Pisa herself had been shattered, and the pre-eminent sea-powers became Venice and Genoa.

It was in maritime societies such as these, where the wealthiest citizens were shippers and merchants, and where, too, a secular education was not difficult to come by, that any new ideas or discoveries in the field of navigation would readily be seized upon and developed. There is no need, therefore, to reject the strong tradition which assigns the invention of the mariner's magnetic compass to Amalfi, even though the story has been embellished with fictitious or erroneous names and dates in its sixteenth-century version. An Italian poet, William of Puglia, when writing the *Gestes* of Guiscard between 1109 and 1111, apostrophizes Amalfi, then at the height of her fame as 'Nauta maris coelique vias aperire peritus'—famous for showing sailors the paths of the sea and sky. And this would exactly apply to the discovery of the use of the magnetic needle which was made about this time.

That the magnet stone could attract iron, and that a piece of iron thus attracted could then in its turn attract other iron had been a familiar fact since ancient days. And it remained a marvel that this dirty brown stone possessed this hidden mysterious power, which was believed to be not only that of drawing iron, but also of discovering an unfaithful wife. Such a stone was naturally part of the stock-in-trade of the magician when he wished to build up a belief in his supernatural powers by displaying mysteries and wonders. St. Augustine relates his astonishment at seeing a magnet-stone pick up a ring, and this ring pick up another, and so on until a chain of rings was formed by unseen force. But he goes on to say that one of his friends had seen an even greater marvel. Fragments of iron laid on a silver plate had been made to move to and fro at the command of a stone held beneath the plate, and so not even touching them. Moreover, as Roger Bacon reminds us, whereas one end of the stone draws iron to itself, the opposite end if then pointed towards it will make it flee away like a lamb from a wolf. And although he was writing at a much later date (A.D. 1267) it is Roger who provides a clue to the way in which the directive property must first have been noticed. He is explaining that the attractive power is still exercised even through water. If a needle thrust into a straw is floated in a basin of water and the magnet stone is held beneath the basin, the needle will dive down to seek it. And if the stone is held above the basin the needle stands erect and rushes towards it. And when at last the stone is taken away the needle turns and stands still with its point towards the north. If as seems probable the magicians were accustomed to show the power of the magnet over such floating fragments of iron, the fact that these always turned into the meridian when left to themselves could not have escaped notice. And the behaviour of little wax models of ships or of other objects floating in a basin was certainly one of the means of divination. In the miniatures illustrating the medieval history of Alexander, his reputed magician-father, Nectanebus, is always represented with his tube for star-gazing, his rod and his basin. In the story as told by Julius Valerius in the late third or early fourth century, the Egyptian kings, of whom Nectanebus was the wisest, used to destroy their enemies, not by fighting but by magic. Foreseeing a hostile attack Nectanebus retired into his palace.

'ibi se solitarium abdebat invecta secum pelvi. Quam dum ex fonte liqui-dissimo impleret, ex cera imitabatur navigii similitudinem effigiesque hominum illic collocabat. Quae omnia cum supernatare coepissent, mox moveri ac vivere visebantur. Adhibebat et iam et virgulam ex ligno ebeni et praecantamina loquebatur, quibus vocaret deos superos inferosque, sicque laborabat pelvi naviculam submergi. Ex quo fiebat, ut simul cum submersione illius cerae et cereis insessoribus, etiam omnes hostes, si qui adesse prae-nuntiabantur, pelago mergerentur.'

Briefly, after filling his basin, Nectanebus made waxen models of a ship and men which he floated in it. They began to move and seemed to be alive, and he approached his ebony wand and uttered the spells with which he invoked the gods, whereupon the ship sank in the basin. He concluded that the approaching enemy ships would sink like the wax ones. But on another occasion for himself the omens were evil and he abandoned his kingdom and departed to Macedonia, where he practised as an astrologer and was consulted by the mother of as yet unborn Alexander.

With the help of a hidden lodestone and iron wire in the wax models such a navy could be made to sail and be wrecked, and the story of Nectanebus must have been invented by someone who had seen or practised such 'magic'. Sooner or later it would be noticed that when the lodestone was removed the models turned north. A trick with a palmed lodestone was actually tried by the famous instrument-maker, George Hartmann, about 1544, when he made a pivoted needle whirl around at his word of command before the Emperor. But by that date the educated layman was not to be deceived. The Emperor smiled.

If knowledge of the polarity of the magnet and of the magnetic needle was indeed an 'escape' from the secrets of the magicians, it becomes easy to understand the odd procedure on ship-board when it first came into use. An account of this is to be found in the writings of two learned Dominicans, Thomas of Cantimpré and Vincent of Beauvais, who were contemporaries. Both friars compiled encyclopaedic volumes *Of the Nature of Things*, rather on the lines of Pliny's *Natural History*, Thomas probably about 1240, and Vincent a few years later. Both write in almost identical terms of the magnet, and must have used a common source. 'When clouds prevent sailors from seeing Sun or star', they said, 'they take a needle and press its point [or rub, it says Vincent] on the magnet stone. Then they transfix it through a straw and place it in a basin of water. The stone is then moved round and round the basin faster and faster, until the needle, which follows it, is whirling swiftly. At this point the stone is suddenly snatched away, and the needle turns its point towards the Stella Maris. From that position it does not move.' (Plate VII.) This certainly sounds like a magic rite, although not all accounts are so dramatic. Jacques de Vitry, for example, who was Bishop of a church in Acre, and was present at the siege of Damietta, had evidently noticed the needle being used during his voyages. 'An iron needle,' he wrote in 1218, 'after it has made contact with the magnet stone, always turns towards the North Star, which stands motionless while the rest revolve, being as it were the axis of the firmament. It [the needle] is therefore a necessity for those travelling by sea.'

There are three still earlier mentions of the sailors' needle than these. Two are to be found in the writings of Alexander Neckam, an English monk who had studied and lectured in Paris University about A.D. 1180. He may therefore have actually seen a primitive compass when crossing the Channel, for in his *De Utensilibus* where he lists a ship's stores he adds : 'They also have a needle placed upon to a dart (*jaculo suppositum*), and it is turned and whirled round until the point of the needle looks north-east [*sic* for north]. And so the sailors know which way to steer when the Cynosura is hidden by clouds.' The pivoted needle was thus perhaps already known, and appears to have been whirled round in the way that the two Dominicans describe, for in his second reference (in his *De ·Naturis Rerum*) Neckam says that after the sailors have 'placed the needle on the magnet stone' it is turned about in a circular fashion, and when it comes to rest it points to the northern quarter. Like several other of the early writers on the subject, Neckam suggests that the needle was only resorted to when the ordinary methods of navigation failed, and this was probably actually the case in the early days of the discovery. 'They do this,' he says, 'when as they are crossing the sea they lose the advantage of the bright Sun during cloudy weather, or when the world is wrapped in darkness at night, so that they do not know in which direction the ship's prow points.'

The third early mention of the magnetic needle occurs in some verses by the satirical poet, Guyot of Provins, which were written about 1205 or perhaps a few years earlier. Here he exclaims, 'O that the Holy Father were unchanging, like the star that never moves! This is the star that the sailors watch whenever they can, for by it they keep course. They call it the Tramontane, and while all the other stars wheel round, this stands fixed and motionless. By the virtue of the magnet-stone they practise an art which cannot lie. Taking this ugly dark stone, to which iron will attach itself of its own accord, they find the right point on it which they touch with a needle. Then they lay the needle in a straw and simply place it in water, where the straw makes it float. Its point then turns exactly to the star. There is never any doubt about it, it will never deceive. When the sea is dark and misty, so that neither star nor Moon can be seen, they put a light beside the needle, and then they know their way. Its point is towards the Star, so that the sailor knows how to steer. It is an art that never fails.'

The poet is clearly aware of some of the finer points of sailors' practice. The needle must be 'touched' to the correct point of the lodestone, in fact its south pole, and the lantern is already in the binnacle, originally the 'habitaculum' or 'bitacle', the small sheltered chamber in which the needle was placed to avoid the wind. It cannot be expected, however, that monks, bishops or poets should observe navigational techniques very closely, or distinguish what they had heard or read about the more primitive methods from what was true of the more advanced. It is certain, nevertheless, that the magnetic needle was in general use, and had long been in use, when Alexander Neckam mentioned it. As to the common story that it had been brought in from China by Arab sailors, there is no evidence whatsoever to support it. A 'south-pointing chariot' known from very ancient times in China has been proved to be no magnetic instrument at all, but a mechanical toy. A south-pointing needle is, however, mentioned between A.D. 1086 and 1093, while a few years later, between 1101 and 1103, its use on shipboard is remarked upon in Chinese annals. If the Arabs knew and used the needle, the fact is not recorded until 1243, while their borrowing a name for the magnetic compass from the Italians should not be overlooked.

Too much, however, must not be argued from mere silence. Men of action were very rarely writers, while scholars and literary men very rarely went to sea, and were still more rarely interested in technical matters. English clerics, of course, were obliged to cross the Channel if they wished to visit Rome, or to attend some foreign court, university or school. But Alexander Neckam was strongly of the opinion that no one should go to sea except under extreme necessity. The fine, strong-looking ship might actually be riddled with borings of teredo, it might strike a rock, or be dismasted in a storm, or perhaps the helmsman would fall asleep! All the same, the Crusades took many people to sea, although they chose a land route whenever they could, crossing, for example, the narrow Strait of Otranto from Bari or Brindisi to Durazzo, and then going on by the old Roman road to

Constantinople, rather than sail direct to Syria by ship. When Ingulphus, Abbot of Croyland, went to the Holy Land in 1064, he travelled overland through Germany to Constantinople, and then across Anatolia to Jerusalem. But in the next spring he was tempted by the presence of a fleet of Genoese ships which arrived at Joppa, and embarked with all his party, 'committing ourselves to the sea', wrote one of them, 'and being tossed with many storms and tempests, at length we arrived at Brindisium, and so with a prosperous journey travelled to Rome'—and thence home across the Alps. Not very long after this English ships were beginning to appear in the Levant. For William of Malmesbury relates how 'King Baldwin of Jerusalem, in 1102, coming out of Assur [Tyre] entered a ship called a Busse, of which the captain was one Goderick, a pirate of the Kingdom of England. They sailed to Jaffa, whereupon the Saracens made towards them with twenty galleys, and twenty ships which they commonly call Cazh [galleasses], seeking to enclose the king's ship. But by God's help the billows of the sea swelling and raging against them, and the king's ship gliding and passing through the waves with an easy and nimble course, he arrived suddenly at Joppa, the enemies frustrated of their purpose.' In fact the Saracens with their low-oared galleys must have been just as nonplussed in face of the lofty broad-beamed sailing buss as Caesar's soldiers were when confronted with the ships of the Veneti. Richard Cœur de Lion sent a whole English fleet to the Holy Land, but he himself avoided the Bay of Biscay, and made his rendezvous with them at Marseilles, although he did not actually meet them until he reached Sicily. Meanwhile the Mediterranean sailors themselves, and particularly those of the Italian maritime city-states, thought nothing of sailing their sea from end to end, and the Black Sea besides. And they built up a body of information about courses and harbours the excellence of which is the best proof possible that they were not only using the sailing needle, but were using it with the wind-rose attached, a device that may well have come from Amalfi in her heyday.

The navigational aids in use at the beginning of the thirteenth century, when Venice had become mistress of the seas (boldly challenged by Genoa), were, as will presently be shown, of a type that demanded an educated man as pilot or shipmaster. They must have been outside the range of the average illiterate sailor, who continued to practise 'with needle and stone', as the phrase went, merely to find the north. But the Italian navies, and the Italian trading and convoying fleets, could find competent men, and it was from Italy that the new methods spread over the Mediterranean basin. As evidence of how they sailed we have a book of sailing directions drawn up about A.D. 1250, and a sea-chart which is dated by experts at about A.D. 1275. And the first point that emerges is that the wind-rose, which alike in the Mediterranean and the north had long been eight-rayed, was now so far developed as to divide the compass or circuit of the horizon into sixty-four points (Fig. 12 and Plate VII). There were still only eight winds with individual names, names that had grown up with the Italian language, and were sufficient for the landsmen. The further fifty-six winds (for a direction or bearing at sea was still a wind) were built up by combinations of these, just as in northern countries the thirty-two winds were built up, but probably at a rather later date, from an original four separate names. The principle was simple enough. Each of the eight was 'a wind'—Tramontane (N.), Ostro (S.), Levante (E.) and Ponente (W.), the four cardinals, with Greco (NE.), Sirocco (SE.), Maestro (NW.) and Libeccio (SW.). In some districts Arab occupation had been long enough to substitute 'Garbino' for 'Libeccio', for El Gharb means the west, and there are occasional dialect variations in the early sailing directions. But from these eight came first of all half-winds, e.g. Tramontane-greco, nor'-nor'-east, and then quarter-winds, e.g. Tramontane quarter Greco for north by east, and finally eighth-winds, e.g. Tramontane eighth Greco. Between each neighbouring pair of winds there was therefore one half-wind, two quarter-winds and four eighth-winds; in other words the sailor was reading his bearings to between 5° and 6°, although, of course, he actually knew nothing of the degree as a unit of 'compass' measure.

It is clear that this refinement of the wind-rose could only have taken place after the discovery of the magnetic needle. It is clear, too, that no such delicate readings would be possible with a needle floating in a basin. Any pilot or shipmaster who possessed a lodestone could set up this latter device when Sun and star were hidden, and could read to 'somewhere between north-east and north', or 'a little north of east', but the finer readings demand that the magnetic needle should have been taken over by the instrument-maker. He alone could balance the needle on a pivot, draw out a geometrically precise wind-rose, and protect the delicately quivering needle by enclosing it in a box. The fact that in some parts of the Mediterranean world the mariner's compass is called 'the box' (*bussola*) rather than 'the needle' (*aguja*) as it is, for example, in Spain, is evidence of the very early date at which this last improvement must have been made. It seems, however, unlikely that the idea of actually attaching the needle to the compass card occurred at once to the early compass-makers. A bare needle is more likely, having the wind-rose underneath it, or marked round the rim of the box. Such bare needles were used for setting pocket sundials in the early fifteenth century, and Leonardo da Vinci, in one of his mystical drawings, shows a large bare-needle compass mounted upon a binnacle (in the modern sense of the word) in a small sailing boat. Whoever it was who first glued magnetized wires beneath the wind-rose or compass card, and pivoted the card itself, must have saved the pilot from many an error of parallax. And it is quite possible that the reason why the English, Dutch and German sailors called the instrument the 'compass' or the 'mariner's compass', or even the 'sailor's compass' or the 'Stella Maris', without mention of the needle, was that they first met it in this form after being accustomed to the mere 'needle and stone'. For the ordinary meaning of the words 'mariner's compass' was simply the division of the horizon into thirty-two points, as contrasted with the astronomer's compass or division into 360°.

The first direct evidence of the subdivision of the eightfold wind-rose by sailors in northern waters is to be found no earlier than A.D. 1240 and comes from an English scholar who was devoting a couple of pages to an attempt to make a drawing of a wind-rose which should combine the twelve winds with which he was familiar from Latin writers with the sixteen winds the names of which he had come across in Norman-French. These latter were the usual combinations of the names of the four cardinals nor(d), su(d), est, ouest, into nor-est, nor-nor-est, su-ouest, su-su-ouest and so on, a terminology hardly distinguishable from its English equivalent. The writer, who was the well-known St. Alban's historian and cartographer Matthew Paris, drew a large circular diagram with sixteen equally spaced rays, writing the Latin and the Gallic (as he called them) names against each. And since the eight secondary winds in the classical system were considered to be 'companions' of the four cardinals, two neighbouring each, he equated them accordingly to the nearest points, Boreas to north-north-east, Caecius to north-north-west and so on. This left the four points north-east, north-west, south-east and south-west as those with only vernacular names. The earliest mention of a thirty-two-point system being used by English sailors occurs in Chaucer's treatise on the astrolabe, that is to say about 1390, but it was no doubt developed considerably earlier than that.

It is a strange and interesting fact that whenever the seaman's compass or wind-rose was drawn or painted it was always given the conventional pattern of a star, that is to say the divisions were not shown by simple radiating lines, but by tapering points like star-rays, and each set of winds, half-winds, and quarters, usually up to sixteen, but sometimes only eight, was in a different colour. The north point was distinguished by an ornament (in later times usually the fleur-de-lis), and the east by a cross, substituted for the initial letter of the wind. The practice no doubt arose from the fact that the needle (as it was supposed) always turned towards the North Star, and did so, as it seemed, by mystic sympathy. And this star was not only precious because it was the sailor's star, the Stella Maris or Star of the Sea, that never moved; it was also by tradition the Star of Mary (Plate VII). The association goes back to the days of St. Jerome, and a tenth-century hymn invokes the Virgin in the words:

> Ave maris stella,
> Dei Mater alma
> Atque semper Virgo,
> Felix coeli porta.

Alexander Neckam, too, apostrophizing Mary as the Rose of Delights and Glory of Womanhood, goes on to call her Star of the Sea and Queen of the Poles. 'Behold the Pole Star!' he continues, 'the apex of the north, shining out on high. The sailor at night directs his course by it, for it stands motionless at the fixed hinge of the turning sky—and Mary is like the Pole Star.'

ponete ilogarbino qnta. Velocto cavo
blanco ecep ro mi ati · ccc·xl·mil px
iozno. Velcapo ecsaline ennalexanoa
a·cccc·lx·mil p garbino ilo x iozno gr
ta. Delocto san beffino arcsautino· olx·
mil p garbino iloponete qita. Velocto
capo ecoga ennacen· ce mil psilacto. uma
co iozno. tra Delcapo ecoga asim· clxxxl
psilacto ii x oi qita. Delcapo ecoga abara
ti. cl·mil psilacto Delocto capo ecoga ab
golfo oara·cxxx mil enf lesiate esilacto
Delcapo ecoga aualanea·cxx mil ple
uate. Velocto capo ec sint anorea enaen
ccxx mil· px iozno ilosilacto qita. De
lcapo sint anorea abarim ecp assaue alo
flume eceine ely mil enf x iozno asila
co. Velocapo sint anorea amargacti enil
psilacto Delcapo ecsco anorea alasgco
losoloino· lxxx mil p leu ite. Velcapo sco
beffino arsautino· oclxx mil· p garbio
ilopote qita. Orac coplito lolibro che se

VI. This page from *Lo Compasso da Navigare*, the earliest surviving Mediterranean Pilot Book, was written in 1296. The first lines read '. . . ponento ver lo garbino quinta. De lo dicto cavo Blanco de Cepri a Damiata ccxl millare per meczo iorno. De lo capo de Saline enn Alexandria. . . .'

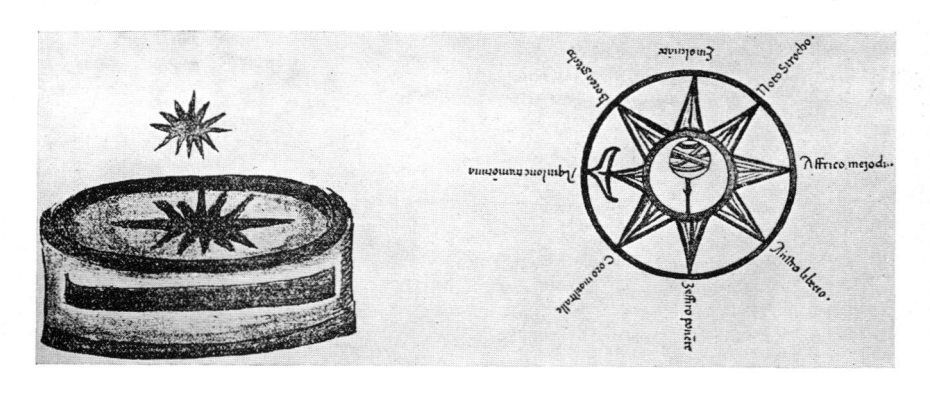

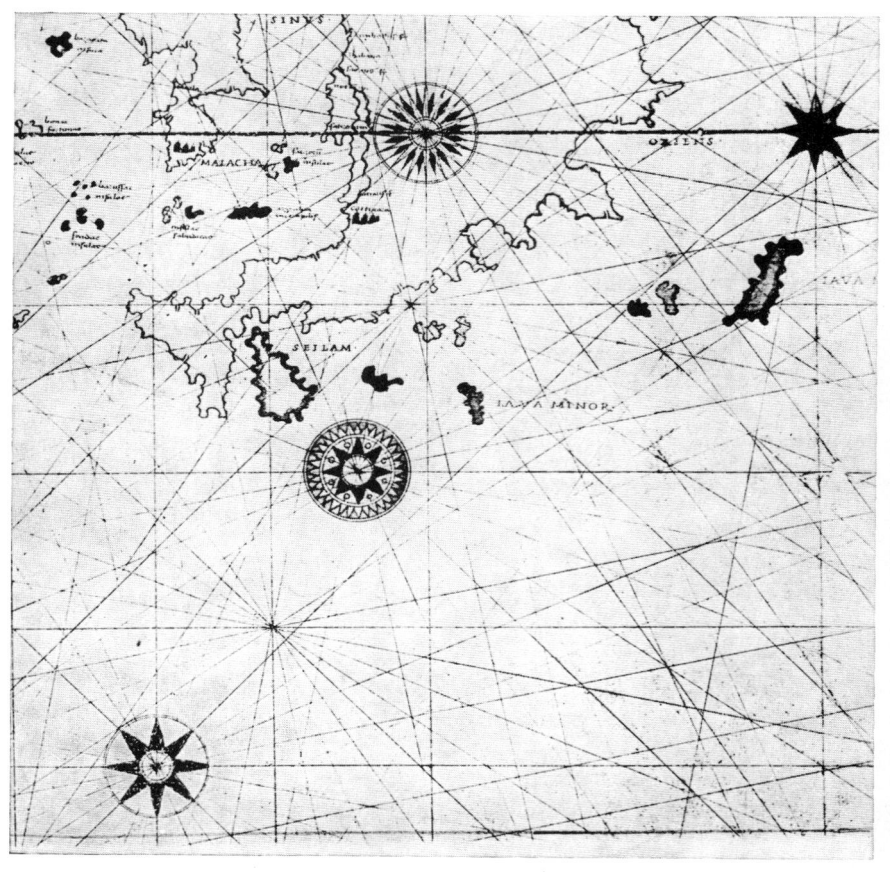

VII. On the compass-box lid is painted the image of the *Stella Maris*, which appears more correctly drawn in the wind-rose and on the chart below. The miniaturist tries to equate classical and vernacular wind names, but transposes *Affrica* and *Austro*. He was painting ornaments on the margin of a cosmographical poem by Gregorio Dati (1363–1435). The chart is by another hand.

But the French scholar, Peter Peregrinus, Roger Bacon's friend when he was in Paris, and the man who wrote in 1269 the first scientific treatise on the magnet, spoke rather condescendingly of the popular belief that the floating needle turned towards the Stella Nautica. The truth was, so he said, that the lodestone carried within itself the likeness of the firmament, and so its poles pointed towards the celestial poles. And it was an error to suppose that the sailor's star was at the north pole, although it circled near to it. Roger Bacon, too, when he was writing for the Pope, took care to call our Pole Star the Stella Nautica, and to point out that it was not the star that attracted the magnet. He was keenly interested in his friend's idea that if only a lodestone could be correctly cut by a gem-cutter and correctly mounted so that its axis exactly matched that of the celestial sphere it would also, once set in motion, revolve daily in exact accord with the heavens. Thus it would provide a perpetual clock, and make all sorts of astronomical observations unnecessary. It would be worth a king's treasury, Roger said. But all Peter's experiments had so far been unsuccessful, and meanwhile he had devised two sorts of magnetic compasses which would help the astronomer to find the meridian and take the azimuths of the heavenly bodies.

The first instrument consisted of a carefully cut and shaped lodestone enclosed in a flat circular box with a tightly fitting lid such as were then made to hold mirrors. A long slender pointer was affixed to the centre of the lid and the sealed box floated in a large basin of water. A thread was laid across the basin to mark the line of the meridian as found by the usual astronomical methods, and when the box swung about and came to rest the pointer was turned to coincide with this line. The box could then be lifted out and marked with the four cardinal points, the quadrants being further divided into 90° as on the astrolabe. With an alidade substituted for the pointer, the instrument would show correct bearings whenever it was placed in water. Centuries later John Dee was to write in the margin of Peter's book that he was wrong in supposing that the needle sought the celestial pole, it turned to the magnetic pole. But in fact this instrument, in which the lodestone was hidden from the observer, was automatically corrected for the magnetic variation by its very method of construction. This was not the case with the second in trument. It

8—H.A.

consisted of a transparent box fitted with an upright spindle turning in cavities in the lid and the bottom of the box. A needle was thrust through the spindle, balanced by a corresponding length of silver wire set at right angles to it. The lid of the box was graduated into four quadrants of 90° each and marked with the cardinal points. Once the needle was magnetized (curiously enough by approaching the lodestone to it from outside the box), the box had to be turned so that the north–south line corresponded with the line of the needle. A sight rule or alidade was fixed to the lid, and by means of this instrument, the designer said, any traveller whether going by land or sea, could find out the direction of his destination, provided that he knew its latitude and longitude. Unfortunately, as Bacon complained, the determination of latitude and longitude had been almost completely neglected in western Christendom, although this was the basis on which maps ought to be drawn. He attempted such a map himself, and sent it to Pope Urban, but unfortunately it has not survived.

It is perhaps because the magnetic needle, after trial, was found to give only a faulty meridian that Peter Peregrinus' work was not followed up, although his manuscript was re-copied and read. But for the sailors, even supposing they had noticed it, the variation was quite immaterial. Their bearings were noted and charted just as they were observed; that is to say they were all bearings from magnetic north. And no trouble occurred until the Great Age of Discovery, with its new mode of sailing by latitude.

Sailing directions, even if only given in the simplest terms, and even if merely handed on verbally from master to mate, must have been continuously in use since sailing began. Octhere gave King Alfred a 'periplus' of the stages of his voyage to the White Sea, and followed it by another of the stages and days' sail to Hetha in the mouth of the Baltic, where Wolstan took up the narration and recited the final stages through the open sea to Trusco, a port near the mouth of the Vistula, but actually on its tributary the Ilsing (Ebling). The course was seven days' and seven nights' sail from the southern end of the Great Belt, and while no bearings are given, the islands and lands lying to 'leerboard', or port, and to starboard respectively, are listed. The next surviving fragment of a 'periplus' is the Bishop of Bremen's summary of the sea route from north-west Germany to Acre which he wrote down a few years after the Norman conquest. Distances when they are mentioned are still in terms of conventional days' and nights' sail, two days and two nights, as for example, from Ribe in Denmark to Zuin in Holland, and thence 'angelosus inter austrum et occidentum' to Prol, perhaps Portland Bill, in England. The Bishop uses only an eight-rayed windrose, but ships move away from Barcelona 'fere versus orientem, declinando tamen parum ad plagam australem', that is to say 'almost due east but declining a little towards south'. The last word must be an error for north, since the ships were bound for Marseilles. From this port they went to Messina in four days and four nights, the course being south-east, while from Messina to Acre was a distance of fourteen days and fourteen nights 'inter orientem et austrum, magis appropriando ad austrum'—southeast but nearer to south.

A very short portulan (as such directions were later termed) of the course from Acre to Venice, dated somewhere about A.D. 1200, gives us little indication of the richness of detail and refinement of bearings that were now accumulating in the notebooks of literate sea-masters, or were stored in the memories of local pilots. The period was one of great splendour in Venice, and mathematical scholarship began to flourish vigorously in Italy after Leonardo of Pisa had written (1202) his book of arithmetic, which introduced the Hindu system of numerals to the west. He wrote, too, on geometry, that is to say on the properties of triangles and their applications to measurement. Himself of merchant class, he had travelled to most parts of the Mediterranean Sea, and had the needs of practical men in mind. Among his most distinguished pupils was Compano da Novara who wrote treatises on astronomy and on the sphere, and became chaplain and physician to the same Pope Urban who commanded an account of his scientific work from Roger Bacon. Roger complained bitterly of the lack of interest in practical mathematics north of the Alps, and of the great expense and difficulty there of procuring instruments for observation and experiment. But the situation was very different in the south. The Norman Kingdom of Sicily (which included southern Italy) was a meeting ground of western scholars with learned

Arabs and Jews who did not disdain the profession of instrument-maker, and here, too, Greek texts were recovered and translated from Arabic into Latin. Towards the latter part of the century the crown of Sicily was worn by Pedro of Aragon, whose brother had become king of Majorca. Their sister was married to Alfonso the Wise of Castile, who fostered the translation of works of Arab scholarship into Latin besides the preparation of new Ephemerides. This king's great compilation, the *Libro de Saber de Astronomio*, did not neglect to describe the making and use of instruments such as furnished the fine Arab observatories in Asia.

It can be little wonder that in a world delighting in the re-discovery of the interest and value of mathematics, in a world too, which, unlike that of the ancient Greeks, held the merchant and trader in high esteem, someone or some group of men would be found to set navigation on a firm mathematical basis. There is no clue to the names of any such men, but a threefold advance took place which could hardly have occurred at random. In the first place, the scattered sailing directions to be found all over the Mediterranean and Black Seas were collected into a single coherent whole. In the second place a scale chart of the corresponding area was drawn for use with the existing advanced type of magnetic compass. And in the third place a method was devised by which a ship-master, sailing with such directions, chart and compass, could work out arithmetically his course made good. All three appear to be Italian in origin.

The comprehensive sailing directions are first found in a book entitled *Lo Compasso da Navigare*, which Italian scholars date as written about A.D. 1250 (Plate VI). In its title the word *Compasso* is used in its original sense of a circuit, for the directions given follow from port to port clockwise round the Mediterranean Sea, beginning at Cape St. Vincent, proceeding eastwards along the north coast, and continuing westwards along the African coast as far as Safi in Morocco. The *Compasso* is written of course in the Italian vernacular, but the oldest surviving manuscript is a copy made by a professional scribe in 1296, and he gave it a Latin *incipit*, calling it *Liber Compassum*. Great care has to be taken in interpreting this word, for it also meant a pair of drawing compasses, or a pair of dividers, and as will presently be seen, it was at one time used as a name for the accompanying sailing chart as well as the written book or portulan.

The unit of distance measurement employed in the *Compasso da Navigare* was what has sometimes been called the 'little sea-mile', which in fact was the geometric mile, with a length five-sixths that of the Roman (or later the Italian) mile. It contained approximately 4100 English feet, and in comparing the distances given in the old sailing directions with those on the modern map the number of miles must be reduced by one-fifth, 60 miles becoming 48, 25 miles 20, and so on. This small mile was familiar to such mathematicians as Compano da Novara, who has already been mentioned, and he gives the length of a degree of the meridian on three scales, namely as 81 geometric miles, 68 Italian miles or $56\frac{2}{3}$ Arab miles. The medieval sailor, however, made no use of the degree for measuring distance, and it was not until the fifteenth century that these three apparently contradictory figures led to confusion—a confusion made worse by the fact that the ancient degree measurement of Eratosthenes was read as 70 miles and that of Ptolemy as $62\frac{1}{2}$ miles!

The use of the magnetic needle, and perhaps also the increasing capacity of ships to sail into the wind, may explain the abandonment of the·old method of defining direction at sea in terms of the following wind; where, for example, Pliny wrote 'from Crete to Carpathus 60 miles with the wind Favonius [the west wind]', the medieval pilot said Carpathos bore '60 miles to the east'. In a set of English sailing directions, however, which will be described later, the two directions, whence and whither, are generally given, so that the course between Crete and Carpathos would be 'west and east', and this may have been used as an intermediate turn of expression in the Mediterranean. In general, however, the language and the type of information supplied in a pilot-book has scarcely altered through the ages. The *Compasso* begins: 'First, from Cape St. Vincent to the mouth of the river of Seville, 150 miles between east and south-east [i.e. east-south-east]. From the said mouth as far as the city of Seville is 60 miles by the river. . . . From the said mouth 5 miles to the south-west is a rock called Peccato which shows above the water. And if you wish to enter the river, beware of the bank called Zizar to the west. And likewise of another bank to the east called Cantara, which is near the cape called Sirocca. In entering the said river by ship you must first take soundings, and note the buoys, and when the water comes in and the tide rises go by the course marked by the buoys . . .'

'From the said Cadiz as far as Talfagar is 30 miles by the south-east. Opposite the said Talfagar in the sea, 7 miles south-west, is a rock. And you must go between the rock and the land, at a distance of $1\frac{1}{2}$ miles from the land. . . . From the said island [Isacaldera] to the Rock of Gibraltar is 8 miles by the south-east and a little east [*per sirocco ver levante poco*]. And on the said Rock is a castle, and under the castle is a good port, with a depth of about 8 fathoms.'

In this last paragraph there is a reminder of the quite natural way in which the subdivision of the winds took place. 'Between east and south-east' came to mean midway between, or east-south-east, while a 'little' east of south-east became a quarter of the way between the winds, the English south-east by east. It will be recalled that Octhere told King Alfred how at Nordkyn he had to wait for a wind that was 'west and a little north'. As might be expected the *Compasso* contains a variety of tentative expressions, such as 'per greco ver levante', from north-east towards east. Or again 'per garbino ver lo ponente terza de vento', a third of a wind from south-west towards west, and even 'per tramontana ver lo greco quinta', which would be N. 9° E., since a 'wind' measured an eighth of the compass. But the quarter-wind is often mentioned, and occasionally the eighth-wind, and these two became the accepted divisions. As the document is a compilation made from various sources, differences in the refinement and naming of bearings (perhaps related to different compass-cards) are only to be expected. There are local differences of dialect too, particularly in the nomenclature of the winds, so that 'mezo di' (midday) often appears instead of 'ostro' for south, while 'garbino' alternates with 'libeccio' for south-west which the Venetians called 'africino'.

Besides bearings and distances the sailing directions gave information about depths and anchorages, the latter accompanied if need be by a note about their security under different winds. In one of the extracts already quoted there was mention, too, of a marked channel, and the *signali* may have been stakes or poles rather than buoys. Sometimes a harbour was closed by a chain (*catena*). For example, 'Marseilles is a good port and has an entry from the west, a depth of 4 fathoms outside the chain. In the entry it is about 14 palms. . . . From Marseilles to the island of Minorca 300 miles to Cape Mahon, south a quarter south-west. And if you wish to pass through the strait between Minorca and Majorca go between south and south-west 340 miles. . . . From Marseilles to the island of St. Peter [off south-west Sardinia] 450 miles, south-east one-third south.'

The landmarks of the principal harbours are given, for example: 'Nice is recognized as follows: within the land are three high mountains covered with snow . . . towards the north-west there is a mountain near the sea called the mountain of Nice.' Genoa harbour was enclosed by a mole, 'and has a cape to the west called Cape Faro on which is a high white tower on which they put a great light at night. The landmarks are as follows: Above Genoa is a pointed mountain called Peraldo, and another forked mountain called the Two Brothers. To the east is a rounded mountain.' Leading marks are frequently named, as for example in entering Pisa (which like Genoa had a lighthouse), 'when the end of the harbour appears between two towers follow that way and you will go clear of all the rocks'. Or in the case of Taranto: 'Put the small island lying to the north-east at mid-poop and the south-west cape of the city at mid-prow and you will go clear.'

Along the shallow western margin of the Adriatic mariners used the sounding lead: 'If you find yourself 20, 30 or 50 miles north-west of Ancona without sighting the port, take soundings [*vada co lo scandallio*] until you are in 7 fathoms, and you will go clear into the said port without sight of land.' Venice was to be recognized by its lofty campanile, and the entrance was by the canal, which had 17 fathoms at high water. Outside the canal mouth the sounding lead brought up mud. It was not often necessary to mention the tides (which are slight) except where there were rocks or banks which were hidden at high water. The course into Acre, the principal port of the Crusaders, was naturally very carefully described: 'The entry to the port is between the said tower [of the Flies] and the rocks. When you approach the port keep at a distance of 4 *prodesi* [cables?] from the city, that is to say from the house of the Templars and the church of St. Andrew, because of the reef beside the said church. And when you have the Constable's house in a direct line with the tower of the Flies you can make straight for the harbour. And when you enter, steer so as to have the city of Cayfas on mid-poop to the east, and the tower of the Flies on mid-prow, and you will reach the harbour clear of the said reef.' The course direct from Acre to Alexandria (480 miles south-west a third west) is given, since it was usual for many pilgrims to visit Egypt either before or after touring the Holy Land. In the old fragment of Venetian sailing directions already mentioned the pilot is directed to sail out due west from Acre for 250 miles and then turn sharp south-west. This detour may have been made as a safeguard from Saracen pirates frequenting the direct route.

Water supply could apparently usually be counted upon, for it is not often mentioned. There was a cistern or reservoir two miles inland near Barca on the coast of Africa, and a little island near Ras el Gibel, which afforded good anchorage at all winds except the north-east, had fresh water close to the sea on its southern shore. There are a few similar entries.

When the circuit of the Sea is completed at Safi, the distance of this little port is said to be 300 miles 'by the north towards the north-east' from Cape St. Vincent 'where we began'. But another section follows, consisting of *Peleio* or long-distance crossings between a number of key-points, especially capes and islands, which must have been of first importance in drawing out the chart of the Mediterranean Sea. In many cases there are sixteen or more bearings and distances from a single point, the distances running up to seven or eight hundred miles. A third section, that on the Black Sea, originated as a separate document from the main *Compasso*, for the scribe writes at the end of the *Peleio*: 'Now the book called the *Compasso da Navigare* is complete . . . and we will describe the Mare Majore [Black Sea].' An interesting detail here appears in the account of Sevastopol. This is said to be a port for winds from east round to north-west, lying under the castle. 'And you drop your bower anchor to a depth of between 20 and 30 fathoms, and your stern anchor is three fathoms.' Evidently, as in Homer's days, ships were anchored with the bows pointing seawards, and according to the modern chart the sharply changing depths quoted are to be found at one or two points near the harbour, for high ground drops steeply into the sea.

The first mention of the chart on shipboard is not until 1270. In that year the French king, St. Louis, set out from his new port of Aigues Mortes with a crusading fleet, intending to make the direct crossing to Tunis which was to be his base. On the way a storm compelled the fleet to take refuge in Cagliari Bay. The king being alarmed, the sailors brought him a chart (which for want of a better word his biographer called a *mappamundi*) and showed him where they were. The true course, according to the *Compasso*, would have been 'da Acque Morte all'isola de San Piero 490 millara per sirocco ver la meczo-di pauco', and then 'de dicta isola a lo capo de Bizerto 140 millare per silocco'. From Bizerta a coasting voyage of 50 to 60 miles would bring the ships to Tunis. The unfortunate king and many of his soldiers soon died at Tunis of disease, but at Michaelmas the English prince Edward, son of Henry III, also took ship from Aigues Mortes. 'And having a merry and prosperous wind', says the chronicler, 'within ten days arrived at Tunis, where he was with great joy received.' However, it took him nearly a week to cross over to Sicily, and he must have been laid up somewhere during the winter, for he did not arrive at his final destination, Acre, until a fortnight after Easter in 1271. His return journey between Acre and Sicily took him seven weeks. The two stages given in the *Compasso* are 820 miles west from Acre to Gozo in Crete, and 720 miles west and an eighth of a wind north-west from Gozo to Cape Passaro in Sicily. When comparing these bearings with a modern chart it must be borne in mind that the variation of the needle was of the order of a quarter east of north (an English 'point') in the eastern Mediterranean in the thirteenth century, so that for west must be read west-north-west in either case. However, the point worth notice here is that these long open sea courses demanded pilots and masters skilled in the use of sailing directions, compass and chart. The early Venetian fragment already mentioned reckoned 14 days and 14 nights between Acre and Sicily—about 110 miles every twenty-four hours.

Of the oldest surviving maritime chart, the Carta Pisana (c. 1275), it can be said at once that it could only have been designed by a man familiar with mathematics as well as with the needs of the navigator (Fig. 12). The two fundamental classes of measurement from which it had been drawn, and by means of which it was to be used, were compass-bearing (in terms of the wind-rose) and distance. The chart had not only to be drawn

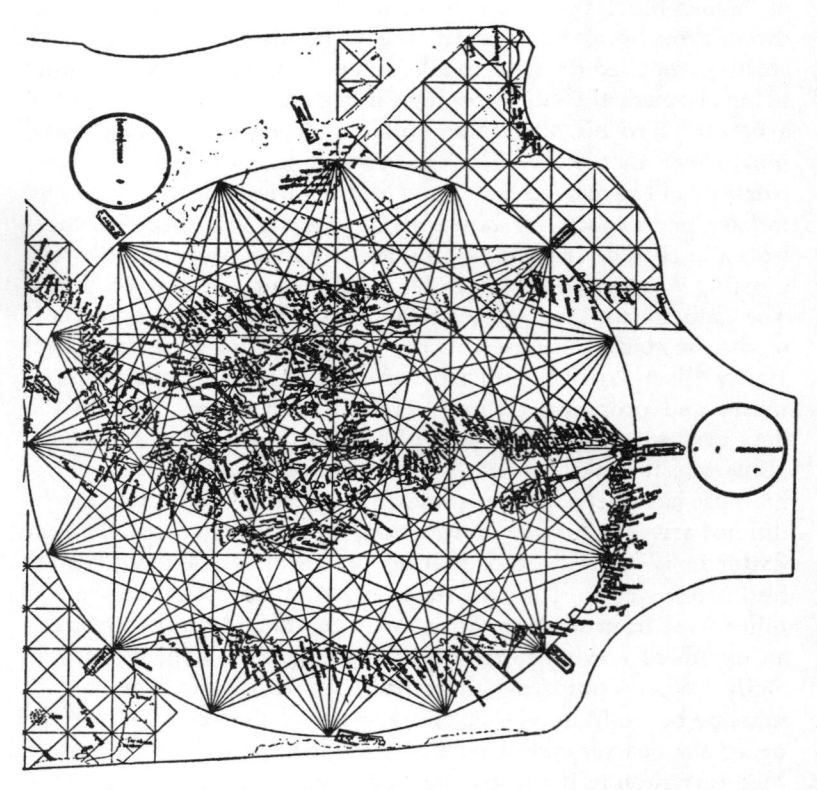

Fig. 12. The sixteen half-winds displayed on the Carta Pisana.

precisely to scale, but to exhibit a scale, and it had also to exhibit the wind-rose in such a way as to facilitate finding the bearing between any two points. The contemporary map-maker when drawing the world paid little heed either to scale or direction. He fitted his map to the size of his parchment, or to some chosen circular, oval or lens-shaped boundary line. Relative size was according to relative importance, or relative knowledge: relative position to a general notion of the four quarters of the world. It is true, of course, that the sea-chart was not the first of all maps to be drawn to scale. The second-century maps of the mathematician Ptolemy, which were drawn on a network of latitude and longitude, were automatically drawn to the scale provided by the measure of the degree of the meridian. But these were astronomers' maps, and were, besides, still quite unknown to Christian Europe, although the Arabs had seen them. A linear scale actually drawn out on the map was seen for the first time on the Mediterranean sea-chart. That on the Carta Pisana showed a length of 200 miles divided into four fifties, and two of these were subdivided into tens and fives. A circle was drawn round the scale with a pair of compasses (perhaps to call attention to it), and it is a rather remarkable fact that the scale was duplicated, one drawn to run vertically, the other horizontally (as is often done today) as though the chart-maker was aware that the parchment might shrink unevenly. This practice was not, however, continued although it is found in a chart of 1311. It follows, of course, that the pilot was now required to furnish himself with the two instruments that always lay to the hand of the practical geometer—hitherto only the architect or master-mason and the surveyor—namely the ruler and pair of dividers or compasses. And he must now, besides, have mastered the first four rules of arithmetic in order to use the scale and to calculate and plot his course.

The manner in which, once and for all, the wind-rose was displayed on the chart, so that bearings could be found without difficulty, also shows remarkable originality and ingenuity. The area of the Mediterranean Sea being oblong, the chart was drawn on a sheepskin, the neck usually to the left or west, although it is to the east in the Carta Pisana. The oblong was divided into two halves and a pair of circles drawn which just touched one another. From the centre of each circle sixteen rays representing the sixteen half-winds were ruled, using a system of different coloured inks so that any particular wind was easily picked out (Fig. 12). From each of the points where these rays met the circumference a quarter of a wind-rose was drawn, consisting of four quarter-winds on each side of the central ray, drawn so as to run to the opposite side of the circle. The same colour system was, of course, used for these secondary wind-roses as for the central ones, and in the later charts some of them were transformed into illuminated stars, presumably matching the 'fly' or wind-rose attached to the magnetic needle or painted on the compass-box (Plate VII). Apart from minor changes, for example in the number of rays drawn from the secondary points, and their extension to the margin of the chart, besides the erasure after use of the circumferences of the two original circles, this network (which was quite independent of the map drawn upon it) remained the standard used down to the seventeenth century (Plate XV).

To find the bearing between his port of departure and his port of destination the pilot laid his ruler between the two, and opening his dividers searched for a rhumb line (as they were called) which was everywhere equidistant from the edge of the ruler, that is to say was parallel to it. If he did not at once recognize what 'wind' this line represented he traced it to its wind-rose, and so to the central rose. Supposing no ray gave him his exact course he took the nearest he could find. Whether these directions as to procedure were written down for the medieval seaman, or taught to him during his apprenticeship, cannot be known. They are to be found in the preface to a book of maritime charts drawn for the English king Henry VIII.

The Carta Pisana is unique in an interesting particular; it shows the technique which the draughtsmen employed in copying outlines correctly. Those parts of the map which lie outside the twin circles (to which the wind-rays are confined) are gridded with small squares and their diagonals, a drawing-office device which is no doubt very ancient, although such lines are usually erased. The coastal outlines of these maritime charts are remarkably accurate when compared with those which were considered acceptable on the world-maps and on the few regional maps of the same period. But even the chart-makers fell back on crude, schematic outlines, as indeed they were obliged to do, once they got beyond the areas of organized Italian trade, for it was upon the trading ships that they relied for the precise data they worked upon. The Carta Pisana, for example, shows only a roughly sketched west coast of Europe beyond Cape St. Vincent, but the charts of a generation later are accurate and detailed as far as the English Channel and Flanders, for the Venetian galleys were now sailing there regularly. South and west Ireland, and parts of Scotland, too, were charted, and it has been suggested that the addition of the Atlantic coasts and Narrow Seas to the chart owed much to the campaign against England, in 1297, of an able Genoese admiral in the service of the king of France.

A distinctive convention, still employed, which goes back to the earliest charts, is the manner of writing place-names. These run inland from the coast at right angles to it, thus leaving its outline clear, and they are written so as to follow one another and be read clockwise round the Mediterranean Sea, in the same order as they occur in the *Compasso*. Different coloured inks were employed to distinguish the degree of importance of the various sea-ports, and in the later charts a flag was painted against each, bearing the emblem or coat of arms of its ruler. This, of course, gave valuable information to a visiting ship. As a rule no inland names or features were shown, but some schools of chart-makers broke this rule and filled the lands with drawings and illuminations. This may, however, only have been when the chart was intended for some wealthy patron. It is such specimens that are more likely to have survived than charts actually taken to sea (Plate XV).

Little or nothing can be said about the way a professional chart-maker organized his business and ordered his workroom. Many charts were signed and dated by the master (the earliest known of these in 1311), but it cannot be assumed that they were the work of a single hand. The arts of lettering and of illumination were taught as separate skills, and this part of the work may have been left to specialist assistants while the master himself dealt with the collation and interpretation of sailing directions and shipmasters' notes, and with the framing of the mathematical base of the chart. Skill with drawing-compass and ruler was required for drawing the network of rhumbs, while the detail of the coastal outlines could only have been derived from pilots' sketch-maps, which had then to be uniformly stylized. One of the most baffling facts about the history of the marine chart is that a single master-copy appears to have been available from the outset, from which all later ones show merely deviations in detail.

The Carta Pisana is so called because it is preserved at Pisa, although it is not impossible that it was made there. But at the beginning of the fourteenth century there was certainly a chart-maker's business at Genoa. The master was Petrus Vesconte, who signed known charts of 1311, 1313, 1318 (two) and 1320, while another member of his family, Perrinus Vesconte, signed charts of dates 1321 and 1327. That so many charts from one business should have survived points to a very large output. An atlas of Vesconte's maps was included in a famous book which Marino Sanudo addressed to the Pope at Avignon in 1320. It comprised, besides nautical charts, a map of the world, a map of Palestine and plans of Jerusalem and Acre, for Sanudo was earnest for a new Crusade. Palestine is drawn on a grid of small squares measuring roughly 3 × 3 geometrical miles.

It was to Genoa that a young man, originally educated for the Church, came with his parents in about 1310, when political troubles drove them from Pavia. The youth was apprenticed to the art of book-illumination with a view to a livelihood, and as in his autobiography he says that 'Genoa showed the Pavian the art of map-making' it is most probable that he met the young men in Vesconte's workshop. Elsewhere Opicinius (for that was his name) disclaimed any professional knowledge of chart-making, but in middle-age, when he had actually taken orders and had a post as *scriptor* at Avignon, he used to make fantastic drawings which embodied beautifully copied charts. And his remarks show that he knew both Genoese and Majorcan models. It is, indeed, sometimes claimed that the Catalans of Majorca were the origina-tors of the marine chart rather than the Italians. Certainly there was a very flourishing and famous school of chart-makers at Majorca during the fourteenth century, in which the same basic conventions were employed as were used in Genoa, Venice, Ancona and elsewhere.

The Catalans, always a great sea-people, had colonized Majorca and then Minorca when the islands were taken from the Arabs by the rulers of Aragon after 1229. With a key position in the western Mediterranean, the Balearic group had previously been a centre for Arab shipping and Arab piracy. The learned Jewish astrono-mers, teachers and instrument-makers who had served the Arabs continued to reside in the islands, and surviving records show that they enjoyed the patronage of the ruling house of Aragon, supplying its members with astrolabes, spheres, time-pieces and other instruments as well as with almanacs, and such works as astronomical and astrological texts. The kings of Aragon, so remarkable for their interest in science, were also overlords of the famous University of Montpellier in Provence, where the teachers also included notable Jewish mathematicians and instrument-makers, while Sicily, too, as will be remembered, had an Aragonese king towards the end of the thirteenth century. And at the turn of the new century the king of Aragon had a powerful Catalan navy whose victorious admiral, Roger Doria, was a Calabrian. But all that can be said with certainty is that, whether Catalans or Italians devised the new navigation with compass and chart, it was equally familiar to both.

An inventory has survived of the equipment of the ship *San Nicola* of Messina which had been seized and sold by Italian pirates in 1294. The Prince of Aragon demanded restitution, and the items named included three charts, one 'cum Cumpasso'. As this *Compasso* was only valued at 10 *grossi* it has been suggested that this was a pair of dividers and not a *Compasso da Navigare*, but the argument appears doubtful. And in addition there were two lodestones (*calamita*) of which one is described as 'cum apparatibus suis'—with its appurtenances—which can only have been the needle and rose of the magnetic compass, for which the clerk had no Latin names. Like the biographer of St. Louis he too calls the charts *mappamundi*, this being the only word he knew for maps. The more experienced Opicinius used the term *mappae maris*, maps of the sea, when speaking of charts.

The Catalan records further show that the king of Aragon bought a 'llivre de navegar' in 1323 which it is tempting to think was a navigating manual (like the modern Admiralty Manual), while in 1373 Prince John ordered a 'Carta de Navegar', a marine chart, which was to show as much of the parts to the west of the Strait of Gibraltar as possible. And when he received it in the New Year of 1374, together with some silver dividers and a drawing compass, he said it was as fine a chart as he had seen, and that the instruments were beautifully made. In 1379 this prince was ordering a true *mappamundi* from Majorca, and here certainly the Catalan Jews were innovators, for it was they who incorporated the scale-chart of the Mediterranean into the general world map, and so set a new standard of accuracy for the whole. But the most revealing entry in these royal records comes in 1381, when the prince concedes the right of establishing a public baths in Majorca to Abraham Cresques, a Jew, described as 'magister mappamundorum et buxolarum', master of maps and magnetic compasses. Here, at any rate, chart-maker and compass-maker were one and the same man, although we do not know if it was the same elsewhere. In the same year the young king of France asked for one of the famous Catalan world maps, and from a later letter of Prince John it appears that this was being made by Cresques, who, as well as his son Jafuda Cresques, is called 'our servant', and 'Jew of our house'. Other orders for maps and tables follow, but when John came to the throne in 1387 we learn that 'the Jew Cresques who was making us a *mappamundi* is dead' and if the map is not finished it is to be handed over to 'a Christian master who is apt at such work'. The beautiful Catalan atlas which was sent to the French court, and is believed to be one by Cresques, has survived, and contains some novel features in the preliminary text which will be considered later.

Just as, at the opening of the thirteenth century, the magnetic needle had become sufficiently familiar to provide the poet with metaphors and similes, so a hundred years later the new art of navigation could be taken by a poet as his theme. This occurred between 1306 and 1313, when Francesco da Barberino devoted thirty lines of his *Documenti d'Amore* to the dangers of the sea and the way that sailors overcame them. The good, experienced mariner, he says, takes some men with him who are learned in the use of the lodestone; he has of course a skilled helmsman, a good 'lookout' and *al Compasso steino* (the painted *Compasso*), which curiously enough, as he explains in a footnote, is the name he uses for the chart: 'It is the chart on which in map-fashion the ports and seas, the distances and dangers and lands are shown.' But besides the lodestone and the chart, the sailor must not forget his *orologio*, his time-piece, which in those days was the sand-glass. It marked out his night-watches, and besides he ought not to pass a single hour without noting how far he had advanced on his course according to the wind that was blowing, and how near he had come to his port or to the rocks. And before trying

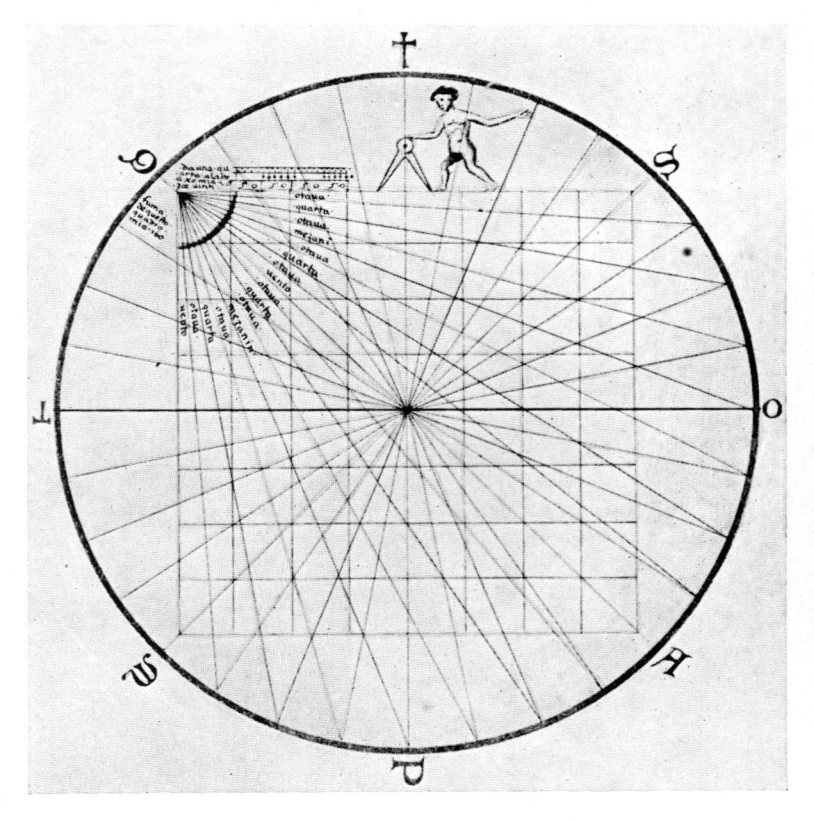

VIII. The 'circle and square' diagram for resolving a traverse graphically, from a MS of 1436. Each 'wind' (45°) is divided into eighth, quarter and half winds. The enlarged scale shows the 20-mile scale divided into halves (50/100), tenths (5/50) and units (half-tenths), allowing any total of miles to be measured. The legend in the left-hand top corner reads 'Each quarter is 20 miles' and 'The whole of this square is 160 miles'.

IX. An astronomer's astrolabe. On the lower plate the sky as seen by an observer is divided by lines of equal altitude and of equal bearing. The upper pierced plate, which rotates, shows the position of leading stars, whose altitudes and azimuths can thus be read once the instrument is set.

to follow how the seaman could use his sand-glass in this way to tell his position it is interesting to notice that an Arab writer, Ibn Khaldûn, who was born in Tunis, notices this new use of the word compass to mean the chart. Writing about 1377 he says that all the countries round the Mediterranean Sea are drawn out on a sheet of parchment in their correct shape and positions, with lines showing the different wind directions, and sailors call this 'folio' a *Kunbâs*, and use it for their voyages. The chart, in fact, did 'compass' the whole circuit of the Mediterranean Sea just as the sailing directions did, and the same name served for both.

The two 'compasses' between them informed the sailor of the course which he should take to get from any one point to any other, but in actual fact he might have to run before a contrary wind in a storm, while even in fair weather he would often be obliged to make a tack or traverse. How then did he actually make his reckoning? It appears that he used tables. In 1382 the king of Aragon ordered payment to be made to Abraham Cresques for 'certain tables' the Jew had provided. And ten years later a new king, the former Prince John, sent to Majorca for the *mappamundi* and navigating tables (*Taules de navegar*) which he had ordered. But it is not until 1428 that we have an actual copy of the tables used, when they appear in a Venetian manuscript written for a dignitary. They also are set out, and not as any novelty, in a navigating manual or instruction book prefixed by Andrea Bianco to a set of his charts. They are collectively termed the *Toleta de Marteloio*, to be used by the Raxon (rule) of Marteloio. Two rather later copies are known, and the four differ among themselves only in respect of a few copyists' errors. As might be expected, they consist of the resolution of certain right-angled triangles, and they contain no actual error greater than a mile. An earlier specific mention of the *Marteloio* (this word has not been explained) occurs in 1390, when a Genoese inventory lists a *martelogium* as belonging to a man who also had a *carta pro navigando*, presumably therefore a pilot. But we are able to put the tables right back to the thirteenth century, and the beginning of the new era of navigation with compass and chart, since their existence is implied by a paragraph in the *Arbor Scientiae* of Ramon Lull and in a chapter elsewhere in his writings. This great Catalan mathematician and alchemist was writing between 1286

9—H.A.

and 1295, using the popular form of question and answer. Question 192 is: How do sailors measure their mileage at sea? And it is answered as follows: 'Sailors consider the four principal winds, namely east, west, south and north, and the other four winds which derive from the first, namely north-east, south-east, south-west and north-west. And they consider the centre of the circle at which the winds make angles. And supposing a ship sails on an east wind 100 miles from the centre, so many miles does she make on the south-east wind (i.e. rhumb-line). And for two hundred miles, twice the number by multiplication. And they know how many miles there are from the end point of each hundred miles east to the corresponding points south-east'. And then he adds: 'Besides this instrument they have the chart, compass (*da navigare*), needle (i.e. magnetic compass) and Stella Maris.'

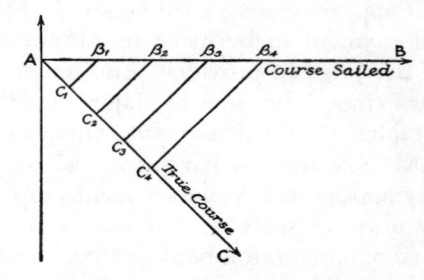

Fig. 13. Ramon Lull's explanation of the course sailed and the course made good.

The 'instrument' (*instrumentum*) can be none other than the table, without which his answer is meaningless. But in fact it is quite simple. Let AB (Fig. 13) be the course sailed along the east rhumb, and divide it into sections AB_1, AB_2, AB_3, etc., of 100, 200, 300, etc., miles. Let AC be the south-eastern rhumb, the angle BAC being 45°. Drop perpendiculars B_1C_1, B_2C_2, etc., from the line AB, then AC_1, AC_2, etc., will be the course made good in direction BC, the lengths of which can be found or calculated by multiplication from the table. The distances B_1C_1, B_2C_2, etc., which are the distances the ship is off her course, are also given in the table or can be calculated by simple multiplication, doubling, trebling and so on. Consulting Andrea Bianco's table, we find that AC_1 is 71 miles, and BC_1 also 71 miles (Cos 45° and sin 45° to radix 10,000 are each 7071). It will be noticed here that there was no question of resolving the course into its northing and easting as in later practice, since medieval sailing had no relation to latitude and longitude. The sole necessary datum was the number of quarter-winds (i.e. the angle) between the true course and the course actually sailed. What particular wind or rhumb was in question was immaterial. Lull's question might have been: 'A ship sails north one hundred miles, how much north-westing has she made?' In his section on navigation in his *Ars Magna* (Chapter 96) he supposes a ship wishing to sail east and obliged to steer south-east. When 4 miles have been covered on this course she has made 3 miles easting and is 3 miles off her true course. This is taking the nearest whole number to 2.83 miles, an approximation very repugnant to modern ideas.

The *Toleta* itself had only eight entries, one for each quarter-wind. It was arranged in three columns. In the first was a list of quarters, one to eight. In the second the course made good (*avançar*), in the third the distance away from the course (*alargar*), assuming that the ship has sailed 100 miles on a course successively one, two, three, etc., quarters from the true course. This table is called the *Suma*. But of course it is not enough for the pilot to know how far he is off his course—he must know how far he must sail to recover it, and at what point along it he will do so. Lull left this matter aside, for he was not writing to instruct sailors. The problem was solved by a second table. Supposing the ship to be 10 miles from her course, then if she sails parallel to (on the same rhumb or wind as) the true course she will never recover it. The table gives for each of the eight quarters that she might turn in from this parallel course, the distance actually sailed (*retorno*) and the distance advanced (*avanço*) on the true course at the point where the two meet (Fig. 14). A third table is provided which is merely derived from the *Suma* by supposing the sailing distance to be 10 miles instead of 100 miles. It gives awkward fractions, but presumably assisted the pilot's calculations.

The earliest manuscript *Toleta*, which is in the British Museum, gives a number of worked examples, but does not deal with the accompanying diagram called the 'circle and square'. (Plate

VIII.) Andrea Bianco explains this as a scale drawing by which the problem of the triangle could be solved graphically. A large circle has the eight winds radiating from the centre, the ends of the rays carrying the initials of the wind-names. The east is at the top. From the north-east point (the left) of the circle a quarter of the wind-rose is drawn as on the network of a marine chart, that is to say with four quarters, or eight eighths on either side of the ray or 'wind' to the centre of the circle. The sixteen rays found on this particular diagram run to the opposite circumference of the circle. A grid of 20-mile squares is drawn within the circle (Plate VIII) and at the top of the diagram is a scale of 40 miles divided into four fifties. Above this again are

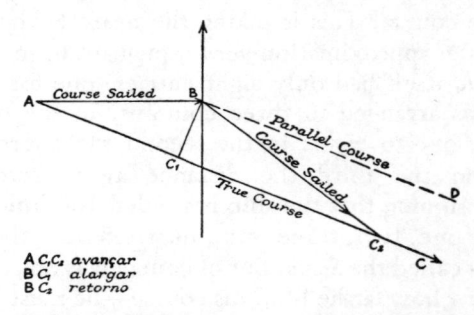

Fig. 14. The Raxon de Marteloio. Supposing AB 100 miles, the tables give AC (avançar), BC (alargar) and BC_2 (retorno) for each rhumb.

scales showing divisions of 2 and 1 miles respectively. Presumably lengths on these scales would coincide with those on the accompanying chart.

These tables must, of course, have been prepared by someone with a knowledge of elementary trigonometry. Such knowledge had long been a possession of the Arab and Jewish scholars of the thirteenth century during which it appears that the new navigation was designed, but by that period it would also have been familiar to Christian mathematicians of the calibre of Leonardo of Pisa and Ramon Lull. However, there is no hint of the assembling of any 'Junta' of mathematicians to prepare the tables such as was called upon to deal with the second navigational revolution in the fifteenth century. Those who used the *Toleta* needed, as

Andrea Bianco said, only to know how to multiply and divide, and it is worth remembering that not only did Leonardo of Pisa write an elementary arithmetic but there was also one written by the Englishman, John Holywood, who was a teacher in Paris during the first part of the critical century. This took the learner up to the Golden Rule or Rule of Three.

It will of course have been noticed that in order to use the *Toleta* the pilot must know how many miles he has sailed at each stage. But all he needed to do was to use his indispensable hour-glass to time his passage on each limb of the traverse, as indeed the poet of the *Documenti d'Amore* pointed out. For every master and pilot prided himself on knowing exactly how much way his ship was making. He knew the ship, he considered the wind, he watched the sails, he watched the water. In fact, it was a matter which just could not be explained to the landsman. A good sailor knew his ship, and that was all.

VI

In the Eastern and Western Oceans

IT is a strange thing that there is no mention, hardly even a hint, of the taking of Sun or star by the medieval Mediterranean seaman in order to check his position. Furnished with the magnetic compass, with chart and sailing book, he seems to have had complete confidence in his dead reckoning, and the working of the Rule of the Marteloio, to bring him to his port. And that port might be as far afield as Galway in Ireland, as Bruges in Flanders, or as the farthest recesses of the Black Sea. Even in the tales told of ships driven off their course—an Italian vessel, for example, which, bound for the Channel, ended up, dismasted and rudderless, in the Lofoten Islands—it is never said that the pilot took the Sun, although from time to time he used the lead-line. No astronomical instrument was as yet carried aboard ship, unless it happened perhaps that some noble or kingly passenger was accompanied by his astrologer. For astrologer and astronomer were all one. And the making and use of instruments was becoming increasingly familiar at the universities, familiar at least to the select few who entered seriously upon the courses of geometry and astronomy which formed part of the advanced studies termed the *quadrivium*. The poet Chaucer was perfectly at home with such operations as finding the meridian by a morning and an evening observation of Sun and shadow, or the determination of latitude by the upper and lower transit of a star. He could tell the time by his astrolabe, or even by the length of his own shadow if nothing else was handy. But such knowledge did not become part of the ordinary sailor's equipment until the Great Age of Discovery. Northern sailors did not even possess the chart, and Chaucer's shipman relied chiefly upon his well-stored memory:

> 'But of his craft, to reckon well his tides,
> His streams and his dangers him besides,

His harbour and his moon, his lodemenage,
There was none such from Hull to Carthage. . . .
He knew well all the havens, as they were
From Gotland to the Cape of Finisterre,
And every creek in Brittany and in Spain.'

Nevertheless, in the Indian Ocean, so it appears, the height of the stars was relied upon as the main guide to position. If we are to believe Nicolo di Conti, an Italian who travelled in the East during the early fifteenth century: 'The Indians navigate chiefly by the Pole Star. According to the greater or lesser height of the pole they measure the direction and distance of places, and in whatever place they are they know only by these measurements.' And this is confirmed by the relation of Marco Polo, a young Venetian who went out to China from a city in which the 'new navigation' was already being practised, and where it may even have had its birth. His father and uncle were in business as merchants. They owned their own ship, and in 1255 sailed in it to Soldaia in the Black Sea, whence they started their first journey overland to the Far East. The elder of the two must have acted as *padrone* or captain of the ship, and although as was customary a master and pilot were aboard, the two owners would have summoned and directed the daily consultation about the ship's course. They would thus become familiar with the methods and the equipment that were used.

While Maffeo and Nicolo Polo were abroad the infant Marco grew up in sea-loving Venice, and it was as a well educated eager youth of seventeen that they found him and took him back to the Grand Khan's court on their second visit, made in 1273. It was three years before they arrived at Peking and the young man, who had meanwhile mastered the Mongol tongue, became a trusted civil servant, travelling widely in the Khan's service. He was once, if not twice, put in command of a small fleet, and voyaged in the China Seas, and when on his return he was commanded to relate the novelties of the voyage to his master, he emphasized the safety of his mode of navigation. It was just then that the question arose as to how best to send a young princess, a prospective bride, away to her groom in Persia. It was strongly argued that, according to Marco's experience, a sea voyage would be safer and quicker than a land journey, and as all the Polos wished after so many years' absence to go home, it was put to the Grand Khan that the knowledge all three 'Latins' had of the *practica da navigar* or art of navigation made them the most suitable escort for the girl. The ruler very reluctantly agreed to let them go, and it was this voyage home that gave Marco Polo the opportunity of commenting on matters of local navigation, and conversing with eminent pilots and navigators of the Indian Ocean. They showed him their maritime charts, and he obtained information from 'writings' that appear to have been in the nature of sailing directions, so that in these respects the Indian pilots must have been as advanced as their opposite numbers in the Mediterranean Sea, and as (according to their own literature) the Arabs also were.

When the Chinese fleet reached the East Indies, where they had to stay some months to await the monsoon, Marco remarked on the fact that the Pole Star could not be seen. This is not really the case in the north-western parts of Sumatra where they were, but it is correct in the sense that even there the star is still too low to be used for purposes of observation. Fra Orderic, who travelled there between 1316 and 1318, remarked on the same point, and Sir John Mandeville writing more than fifty years later develops it further. 'Ye shall understand', he says, 'that in this land [Sumatra] and in many other thereabout, men may not see the star that is called Polus Arcticus, which stands even [due] north, and stirs never, by which shipmen are led, for it is not seen in the south. But there is another star which is called Antarctic and that is even against [i.e. opposite] the other star, and by that star are shipmen led there, as shipmen are led here by Polus Arcticus.' But Marco Polo himself does not say what star replaced the Stella Maris, for in fact there is no star exactly at the South Pole. The brilliant Canopus suggests itself, but it would not be visible all through the year, and the Southern Cross was perhaps used as well. It was Canopus that Mandeville (borrowing from Pliny) described as shining over Ceylon, for Pliny had related how some envoys from that island had come in his day to Rome: 'These men marvelled at the new aspect of the heavens visible in our country, with the Great and Little Bear and the Pleiades [all sailors' stars]. And they told us that in their own country . . . Canopus, a large and brilliant star, lights them at night.'

According to Marco Polo, however, the use of the northern Pole Star began at Cape Comorin, where, he said, it can just be seen. But as he goes on to say that there it stands one cubit above the horizon, his expression 'just be seen' must clearly be taken as 'just be observed' astronomically. For Cape Comorin is in Lat. N. 8° 5', and when in later days English sailors were observing star altitudes they used to say that they 'lost the pole' in somewhere about that latitude. Marco goes on to give two further pole heights, namely one of 2 fathoms (brassi) at Malabar, and one of 6 fathoms at Guzerat. It is impossible, however, to translate these linear measurements into any scale of degrees. A cubit in Strabo's day was the equivalent of only 2°, so that a fathom would be 8° and 6 fathoms 48°. But the fact that Marco Polo had received these measurements in terms of a foreign unit, and that he dictated his memoirs long afterwards when a prisoner of war to a man who wrote them down in a French-Italian dialect, and furthermore that we only have his text in translation and retranslation, makes it almost inevitable that any accuracy that the figures had has disappeared. What remains, however, is of sufficient interest, namely the actual use of star altitude to determine position. Other figures have suffered in the same way. Marco learned for example that the great island of Madagascar where the roc lived was a thousand miles south-south-west of Socotra, but actually such a course and distance would bring a ship no farther than Mombasa. Nevertheless, the same informant correctly told him of the strong southerly current between Madagascar and the mainland which was the effective deterrent to any advance farther south by the Arab and Indian traders on the East African coast at that period.

The Arab navigation technique, as Vasco da Gama learned when he sailed round Africa, certainly included the observation of the stars. In the days of primitive sailing a Sanskrit document had described the *Mu'allim* or pilot in the following terms: 'He knows the course of the stars and can always orient himself. He knows the value of signs, both regular, accidental and abnormal, [of] good and bad weather. He distinguishes the regions of the ocean by the fish, by the colour of the water and the nature of the bottom, by the birds, the mountains [i.e. landmarks] and other indications.' Such were still the means of navigation in use when Cosmas Indicopleustes traded to Ceylon. But after the days of Mahomet more was demanded. For it was written in the Koran: 'He [Allah] it is who hath appointed for you the stars, that ye guide yourselves thereby in the darkness of land and sea. We have made the signs distinct for the people that have knowledge.' And there was no lack of Arab astronomers with knowledge to chart the heavens. Indeed Ibn Mâdjid whom Vasco da Gama met at Malindi was 'Mu'allim Kanaka' or pilot-astrologer, like his father and grandfather before him, and the author of the nautical instructions and rutters known as *Al Muhêt*, 1468–89.

Nevertheless, there must have been many ignorant shipmasters afloat who still relied on the traditional signs, and who even neglected the magnetic needle. Or so we must judge from travellers' tales. The narrative survives, for example, of a man, presumably a pilgrim, who had taken ship on the Red Sea for Jiddah. 'In the late evening', he says, 'we were rejoicing in the sight of birds circling above us from al-Hejaz.' Land was not in sight, and a storm sprang up: 'We did not recognize the direction we were making for until some stars appeared, and some indications could be got from them'. The sky cleared the next morning, and the passengers could see the mountains of the Hejaz. The captain, too, must have picked up his landmarks, for he told them they were only two days from Jiddah. This was in terms of the familiar 'day's journey' by land, which was a unit of from twenty to twenty-five miles. A contemporary writer Al Marwazi gave the 'day's sail' as 150 miles or at about 6 knots, a much higher estimate than the figures, equivalent to a speed of 5 knots or even less, which are given by the Greeks, and have been suggested for the Viking ships.

Both Nicolo di Conti and the famous map-maker Fra Mauro surprise us by stating definitely that in the Indian seas men sailed without the magnetic compass. This was said in the fifteenth century, and there is some equally puzzling information in an Arab manuscript of 1282 entitled *Book of the Merchants' Treasure, treating of the Knowledge of Stones*. This contains a section dealing with the lodestone, in which the author writes: 'Sea-captains of Syria, when the night is dark and they cannot see the stars which show them the four cardinal points, take a vessel of water which they shelter from the wind by going below. They take a needle which they thrust into a [piece of] acacia or a straw so that it forms a cross. They throw it into the water. The captains then take a lodestone of a size to fill the hand or smaller. They bring it towards the surface of the water and make a circular movement from the right with the hand: the needle follows it round. Then they abruptly withdraw it, and the needle turns to stand in the north-south line. This operation', continues the author, 'I saw myself on a voyage from Tripoli in Syria to Alexandria in 1242–3.'

Now this description so strongly recalls the account of the magnetic needle found in the works of Thomas of Cantimpré and Vincent of Beauvais that it seems as if the Arab author has simply translated the Latin of one of these writers into Arabic. And indeed he must have read one or another of them, or else knew the common source which they themselves had used, for the turns of expression are so close. But in fact there are one or two touches that are not to be found in the books of the two Dominicans. The Arab says, for example, that the vessel of water is taken below, and he remarks on the small size of the lodestone, fitting in the hand, so that the best explanation is that he had indeed seen the strange procedure (with its hint of magic) on his voyage to Alexandria, but had since read the description of it and borrowed his words from the writer. The point to notice, however, is that whatever may have been the nationality of the Syrian captains, they were still using the primitive method of floating the needle although, as has already been shown, the precise bearings to be found in the *Compasso da Navigare* indicate that the Mediterranean instrument-makers were by 1242 or thereabouts making a pivoted and possibly also boxed needle for use with a thirty-two-point wind-rose. Even if such an advanced type of magnetic compass were only being used in the more important Italian and Catalan ships it could not have remained a secret from the Saracen world, since piracy was rife and no ship or ship's equipment secure.

That the Arabs knew and used the magnetic needle in some form or other is impossible to doubt, in spite of the positive statements of Nicolo di Conti and Fra Mauro, statements repeated even as late as 1497–8 by a Florentine gentleman who accompanied Vasco da Gama into the Indian Ocean and met the Arab pilots there. What Nicolo had actually said was this: 'In those parts of India [the East Indies] navigation is carried on by means of the stars of the Antarctic or South Pole, for the stars of our North Pole are scarcely visible. They do not navigate by the needle, but conduct their navigation according as they find the Pole Star high or low. And that they do by certain means.' Leaving aside for a moment the nature of the 'certain means', it appears that what he is contrasting is the Arab and the Italian ways respectively of determining and resolving the 'course made good'. The Mediterranean pilot never took his eyes off the needle, and based upon it his orders to the steersman, while he worked out his course entirely by his records of bearing and distance. The Arab might use the needle to check his orientation, but he determined his position relative to his port of destination by taking a star-sight. It was for 'navigation' in its narrower sense and not in its general sense that the magnetic compass was not used. That explains the apparent contradiction.

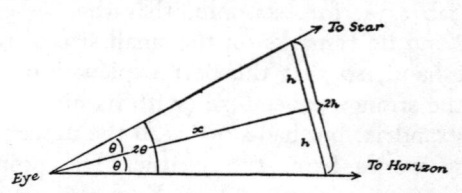

Fig. 15. The principle of kamal and cross-staff. The angle is defined by the ratio of h to x and twice this angle is the distance of the star to the horizon.

What those 'certain means' were which the Arabs employed for finding the altitude of a star only became clear to westerners after the discovery of the sea-route to India. But there is no doubt they were not then new, and they have remained in use almost to the present day. Observations were in fact made by a very simple and ingenious instrument called the *kamal*, which depends upon the familiar principle that an object of fixed length will measure the height of any heavenly body above the horizon according to the distance at which it is held from the eye. In Fig. 15, if the angle subtended at the eye between the horizon and the star is 2θ, and the fixed length is $2h$, then the length x

is $h \cot \theta$; if it is $2\theta'$ the length x' is $h \cot \theta'$ and so on. Since the tangents and cotangents of all angles are known and h is known, any angle can be expressed in terms of the length x. In the case of the *kamal*, a small wooden tablet represented the length $2h$, and a string through the centre of the tablet represented the line x. The observer held up the tablet so that it just covered the space between the horizon and the star (or equally he could use it for measuring the space between two stars). Meanwhile he held the string in his mouth, and shortened it until it was taut, i.e. to length x (Plate XI).

In the simplest instruments the user had the known star altitudes for each port on his route knotted on the string and recognized them according to each length found. But more usually the string was knotted at distances corresponding to *isbas* of 1° 36', each four of which make a *dubban* of 6° 24'. The *isba*, it has been suggested, corresponded to the angle covered by a finger held at arm's length, the *dubban* to the angle covered similarly by a palm. There seems to be no doubt that hand and body measurements were employed in different parts of the world to get approximate star and Sun heights. There was, for example, an Icelandic priest who went to the Holy Land in A.D. 1150, and on his return reported that 'by Jordan, if a man lies flat on the ground, raises his knee, places his fist upon it, and then raises his thumb from his fist he sees the Pole Star just so high and no higher'. And such expressions occur as 'when the Sun is shaft-high' to define the time of day, or 'the star was a man's height above the horizon' to define latitude, and provide further tantalizing puzzles, since there is no mention of the distance at which each 'yard-stick' stood. The Arab *kamal* was constructed on the same principle as the cross-staff, which was first described in Latin by a Provençal Jew in 1342, but this instrument does not come into the navigator's story until the Great Age of Discovery. It has to be assumed that in the case of the *kamal* allowance was made in knotting the string for the fact that it was drawn aside into the mouth instead of running, as the geometrical figure demands, straight to the eye (Fig. 15). Nor must it be forgotten that the Pole Star was several degrees from the celestial pole, and yet there is no indication of how the correction for this was made. Even Roger Bacon, who knew perfectly well that the star revolved, spoke of it as 'at the very pole' when describing how the Arab astronomers used it to measure the length of a degree of the meridian and the size of the Earth. Twice in its revolution, of course, the Stella Maris does stand at the altitude of the celestial pole, and these positions can be recognized by noting the angle at which two other stars lie, namely the stars in the Lesser Bear known as the Two Brothers, or the Guards. A rule based on this relationship was taught to sailors in the late fifteenth century, and will be considered in its place. But it is impossible to say whether the medieval Arab sailor knew it, although it seems probable.

That so very little can be said with any certainty about early navigational practices is in part due, as has already been said, to the very simple fact that the vast majority of writers were landsmen, who belonged besides to the class that never engaged in any manual or practical work. Technical processes were therefore of no interest to them, and when they wrote about the sea it was rather to deter men from the rashness of going aboard a ship. According to an Arab writer: 'The Sea of Darkness [the Atlantic Ocean] is boundless, so that ships dared not venture out of sight of land: for even if the sailors knew the direction of the winds, they could not know whither those winds would carry them, and as there was no inhabited country beyond, they would risk being lost in mists and fogs.' That is a literary man's point of view. He conceives of sailors as adventuring about the seas with no particular objective, and as deterred from this voyage or that by broad philosophical concepts such as that there is only a single habitable land-mass with a circumambient ocean. Actually a sailor knows nothing of such matters, but goes to sea like a man to his office, along a set route for a set purpose, his livelihood. And this had long ago led the Arabs as far as China. That they did not sail the Atlantic, except along the West African coast, where there were valuable fisheries, was because there was no reason to, not that they 'dared not venture'.

Among the Arabs, and indeed among all sailors, there were always to be found the more primitive and the more advanced, in respect of methods and equipment. A seventeenth-century writer, Abbé Fournier, who had made a great study of navigation, said that the floating needle was then still in use in northern waters, and in an English political poem of about 1436 the writer speaks of our sailors as voyaging to Iceland 'by needle and stone', which certainly sounds like the old-fashioned way. Fra Mauro, too, put this note on his great *mappamundi* of 1458 about the East Sea, presumably the Baltic: 'Per questo mar non so navigar cum carta ni bussola ma cum scandio'. That is to say 'navigation is not by chart and magnetic compass, but by the sounding lead'. And if, once more, 'navigation' is taken in its stricter sense, this is true. The compass might be used for orientation, but the course was found by taking soundings, and not in the Mediterranean way by taking bearings.

There was good reason for this, for whereas the Mediterranean Sea consists of a series of deep basins, the seas of north-west Europe are shallow seas. And because they are shallow—forming an enormous extension of the continental shelf—they are subjected to tides of exceptional range, and to tidal streams sweeping now this way, now that. As a result of the mutual interference of river and tidal streams there are deposits of river silt and longshore sand and shingle which are always shifting their shape and position. Depths are obviously of first importance, and the changes of depth twice daily with the tides. As he casts the lead the seaman learns also the nature of the 'grounds', the sand or shell or mud which he brings to the surface from the bottom of the sea. And this navigation with the lead, by depths and grounds, has the happy advantage that it needs no light, and can be carried on in fog and darkness. Chaucer's shipman, it will be recalled, had no peer in his knowledge of tides and Moon, of 'streams' (currents) and dangers, and of lodemenage or pilotage, which is essentially the use of the lead, and of leading marks, as to which he knew every creek and haven on the Atlantic coast. He must have had his 'needle and stone', but it is not mentioned.

And when we examine the first English pilot-book it proves to be completely different from the Italian *Compasso*, having all the emphasis on tides, depths and grounds, with distances scarcely mentioned. The English called the pilot-book a *rutter*, a word borrowed directly from the French *routier*, a book showing the route or course. The oldest surviving English specimen, however, bears no name or title, for it is only known through a copy made by a professional scribe of the reign of Edward IV who was compiling a book of old documents for a noble patron. The rutter, or parts of it, may well go back to the fourteenth century, for it is a compilation including several sections, although the scribe does not distinguish them. The first part comprises sailing directions for the English coast from Berwick round to Land's End and Scilly, including the chief Channel crossings. This is followed by a detailed description of the Channel entry, with particular respect to Ushant. The third section ópens: 'In Spain and Gibraltar, this is the course and the tide . . .', and it includes directions for the coast between Brittany and the Strait of Marocco [i.e. Gibraltar], concluding with a number of bearings from Cape Finisterre to Ireland and the Scillies, or Sorlings as they were called. A fourth section is entitled 'A new course and tide between England and Ireland', which covers all the coast of Ireland together with the opposite parts of the Irish Sea and St. George's Channel. Finally there is a section beginning: 'Here be the grounds of England, Brittany and Scilly. . . '.

The variety of bottoms or grounds recognized is quite remarkable, as a few quotations will show: 'In Belle Isle there is in 60 fathoms or 70, small dial sand [that is to say the sort of fine sand used in an hour-glass]. Upon off Penmarche there is in 50 fathoms black ooze. Upon the same in 60 fathoms there is sandy ooze and black fishey stones among. Upon off Ushant in 50 or 60 fathoms there is red sand and black stones and white shells among. . . . Upon Lizard there is great stones as it were beans, and it is ragged stone. . . . Upon Portland there is fair white sand and 24 fathoms with red shells therein. And in 14 or 16 fathoms there is rocky ground and in some place there is fair clay ground.'

Bearings are given in terms of a thirty-two-point wind-rose, and usually, although not always, the back and ahead names of the bearings are given. For example, the description of the crossing between south-east Ireland and the Pembroke coast runs: 'Tuskar and the Ramsey, east and west. The Tower of Waterford and Grasholm west and by north, east and by south. And beware the rock men calleth Sampson, for he lieth at the south point of St. David's Head. And keep more near the island than the mainland till ye be past the point and through the sand [sound?]. Then go north till ye come at another rock. And for cause of that rock ye must go north and by west or north and by east for north is even with the rock. And the name of the rock is called the Kep, and he lieth under the water. But the water breaketh upon him, and the breach showeth.'

Leading marks are given from time to time, as for example: 'And if ye go to Chester ye shall go from the Skerries [north of Angelsey] till ye come anent the Castle of Rhyddlan. Your course is west-south-west and east-north-east. And take your sight on the mainland of Wales, Rhyddlan, and the Red Bank in Chester Water, north and south.' (The course should actually have been west-north-west and east-south-east but the variation must be remembered.) The approach to the Thames from the north runs as follows: 'If ye go out of Orwell Wains to the Ness [Naze] ye must go south-west. From Ness to the marks of the Spits your course is west-south-west. And it floweth [i.e. the tide] south and by east. Bring your marks together that the parish steeple be out by east the Abbey of St. Osyths. Then go your course on the Spits south till ye come to 10 fathoms or 12. Then go your course with the Horseshoe south-south-west. And if it be on flood come not by in 8 fathoms. And that shall bring you 11 fathoms then go your course in to Thames with the Green Bank west-south-west.'

There are intricate directions for passing the Straits of Dover including the following: 'If ye be bound to Calais haven and ride in the Downs, and the wind be west-south-west, ye must rere [raise anchor] at a north-north-east Moon, and get you into your marks, the steeple into the fan [?], then go your course east-south-east over, and after your wind and your tide serve your course. And look ye seek Calais at a south-south-east Moon or close at south and by east. And if ye turn in the Downs go not near Goodwin than 9 fathoms, nor not near the Breaks than 5 fathoms.' The references to the Moon are to be understood in relation to the universal method used among sailors of describing the tides. This was in terms of the rhumb or point on which the new Moon stood when the tide was full, and the opposite rhumb on which it would stand (approximately) at the time of the second high tide twelve hours later. Thus 'high tide Moon south-west–north-east' meant that at a particular port the tide was full three hours after the Moon had crossed the meridian, for it was assumed (in spite of the actual obliquity of the horizon) that the Moon passed at equal time-intervals round the thirty-two points of the wind-rose or compass card. Each point was therefore equivalent to 45 minutes, and this figure was also accepted as measuring the daily retardation of the tide. Hence when the Moon was four days old, a place with 'high Moon south-west–north-east' would have high tide when the Moon was west–east. Thus the seaman had to carry in his memory the age of the Moon, unless he could understand and afford to buy a calendar, which explains why Chaucer's shipman had to know his Moon. When in the passage cited above the sailor is told to go to Calais 'at a south-south-east Moon' this would be understood as a reference to the Moon when new, and the retardation made according to its age. In the fifteenth century, and perhaps earlier, there were eight standard circular diagrams in use covering eight pairs of rhumbs—sixteen in all—corresponding to a south–north Moon, a south-west–north-east Moon and so on, which made the determination of the hours of high tide, half tide and low tide a simple matter. These will be referred to in a later chapter. (Plate X.)

Returning to the English rutter, we find such information as 'a south Moon maketh high water within Wight', and 'all the havens be full at a west-south-west Moon between the Start and the Lizard', while 'From the Foreland of Fontines [the Bec du Raz] to the Strait of Morocco a south-west Moon maketh highest water.' Portuguese pilots used to express this last statement as 'Moon north-east south-west, full sea'. It was not, however, merely the rise and fall of the tides that the seaman had to be familiar with, but the alternating run of the tidal streams. In the sailing directions for the inner waters of the Bristol Channel, for example, he is warned 'Beware of Iron Grounds, and of your streams of flood, for they sit north-east on the Iron Grounds. And on ebb spare not to go, for the streams of Bridgewater sit west-north-west. And beware of Colum Sand. It floweth from Lundy to the Holmes east and west, and from the Holmes to go clean of the Wash Grounds and of Langbord the course is north. An[d] ye come on ebb, and sith go east-north-east with Portishead, but if ye have a quarter tide at the Flat Holme ye may go east-north-east or east and by south, and go over Langbord with Kettleswood with a good ship, for ye shall have 3 fathoms on the sand or more by that ye come there.'

In the Mediterranean *Compasso* it was only in the entry to such a river port as Seville, or into the canals of Venice that the rise and fall of the tide had to be considered, and there was only one formidable tidal stream. This was in the narrows of the Strait of Messina, where the horrid Scylla had snatched Odysseus' men, while Charybdis sucked and gurgled under the opposite cliff. A brief paragraph describing the run of this stream is included in the same collection of Venetian papers on shipping and navigation in which the Rule of the Marteloio first appears. The Rule of the Current (as it is termed) states: 'In the first place wherever there is an east Moon the current of water flows past Faro [on the mainland] towards the south. When the Moon is south the current turns to the north, and when the Moon is west the water turns and runs south for six hours. Every six hours the current changes from north to south. . . .'

The long-distance bearings and miles termed *Peleio* in the *Compasso* have their counterparts in the English rutter, although they are bearings only, taken from key-points, and there is no mention of distances. From Cape Finisterre, for example, there are bearings on Mizen Head and Cape Clear in Ireland, the first given as north by west, south by east, the second as north–south. Back bearings from Cape Clear include Ortinger, south by east, north by west, Santoña, south-south-east and north-north-west, Bocco of Bayonne south-east and north-west, Sein (Bec du Raz) east-south-east and west-north-west. A number of bearings from the Forne, a rock off Brittany which is now called the Four, include Falmouth north, Rame Head north-north-east, Abbotsbury north-east, and the Needles north-east and by east.

It is perhaps worth emphasizing that the run from Spain to the British Isles was made directly; there was no question of 'hugging the shore', which in fact was to be dreaded. 'An (when) ye come out of Spain', says the rutter, 'and ye be at Cape Finisterre, go your course north-north-east. An you guess you 2 parts over the sea and be bound into Severn, ye must go north by east till ye come into soundings.' The pilot would, in fact, find the unusually steep edge of the continental shelf about a hundred miles west of Penmarch Pt. in Brittany. 'An if ye have 100 fathoms deep or else 80, then ye shall go north until ye sound again in 72 fathoms in fair grey sand. And that is the ridge that lieth between Cape Clear and Scilly. Then go north until ye come into sounding of

ooze, and then go your course east-north-east or else east and by north, and ye shall not fail much of Steeplehorde. He riseth all round, as it were a copped hill.' A ship bound for the Channel kept its course north-north-east and by north from Finisterre until it reached the 100-fathom line, which would bring it into the latitude of the dreaded Ushant, but well away to the west. Only then was the course changed to north-east, 'till ye come into 80 fathoms, and if it is streamy ground it is between Ushant and Scilly in the entrance of the Channel'. A present-day book of *Navigational Notes and Cautions* for sailing ships says under Ushant: 'Rocks are numerous and far from the land: fogs and thick weather are common: the tidal streams are strong and the extent of their influence seawards undetermined. In thick weather the island should be given a wide berth and continuous sounding be resorted to, and a depth maintained of 60 fathoms, allowing for the rise of the tide. In settled fine weather Ushant may however be rounded at a distance of 10 miles.'

While the monthly cycle of the tides, including the recurrence of springs and neaps, was clearly described by the Venerable Bede, and was familiar to the writer of the old Norse book, *The King's Mirror*, it is not until the earlier part of the thirteenth century that we come across a tide-table for a particular port. This is for 'Flod at London Brigge' and is found, unfortunately without any comment, among some manuscripts written at St. Albans and associated with Matthew Paris, who was very interested in geography and kindred matters. The table consists of three columns, the Age of the Moon (one to thirty days), the corresponding time of high tide (in hours and minutes) and the number of hours and minutes during which the Moon shines each night. At the new Moon high tide is said to be at 3 hours 48 minutes, on the second day 4 hours 36 minutes, on the third 5 hours 24 minutes, and so on. It is in fact mechanically built up from a single observation according to the rule accepted by astronomers that the daily retardation was 48 minutes. It is a scholar's, not a sailor's table. When John Flamsteed drew up his tide-table for London Bridge in 1676 he found the figure highly variable, the retardation sometimes under 30 minutes, sometimes an hour or more, while the difference between the hour of high tide and the Moon's passage across the meridian was only two hours or less instead of 3 hours 48 minutes. This difference may have been partly due to natural and man-made changes in the River Thames, but it is also the case that theory rather than observation was still the rule in the learned world of Matthew Paris' day.

Perhaps of greater interest is the tide-table, again the first of its kind that is known, which is found in the great Catalan Atlas of about 1375 which was sent by the king of Aragon to the French court. This atlas, it will be recalled, was most probably made by the king's chart-maker and compass-maker Abraham Cresques, who had every opportunity for gathering from seamen correct information about the tides at the various ports they visited. By this date it was very usual for astronomers to make use of circular volvelles to show the cyclical changes of the heavens—the phases of the Moon for example—and Cresques chose a circular diagram to show the tides at fourteen places in Brittany and the English Channel (Fig. 16). The circle was in the first instance a wind-rose, foundation of all the sailor's art, and in this case sixteen rays were drawn from the centre, of which the eight representing the separately named winds bore their initial letters. As was already the custom, however, a cross was substituted for the L of Levante (east), and a flower ornament (not yet a *fleur-de-lis*) indicated the north. Besides showing the bearings of the Moon, the circumference of the circle represented twenty-four hours or a complete revolution of the sky, as well as a lunar month, and round it the daily retardation of the tide according to the Moon's age could be counted by halving the sixteen wind points, which were $1\frac{1}{2}$ hours apart. In this particular diagram fourteen concentric circles were drawn, separated into two groups of seven, and upon each was written the name of the port it represented. Then at the correct 'Moons' for each port what was in effect the hour of each of the two high and low waters occurring in a 'natural' day when it was new Moon was marked by the letters P for full and B for low water. The Breton data included those for Sayne (Raz de Sein), Samae (St. Matthieu) and Forn d'Artus (the Four rock), while among English names are Portland, Wight, Winchelsea and Romsey. Above the diagram was written the statement: 'This is the course of the tides. From the Rock of Gibraltar to Penmarch, which is in Brittany, Moon north-east–south-west full sea, and north-west–south-east low water.' One would like to believe that Chaucer saw the Catalan Atlas when he visited the French Court, for he placed his story of the lady who would yield her virtue only if the rocks disappeared precisely on Penmarch Point.

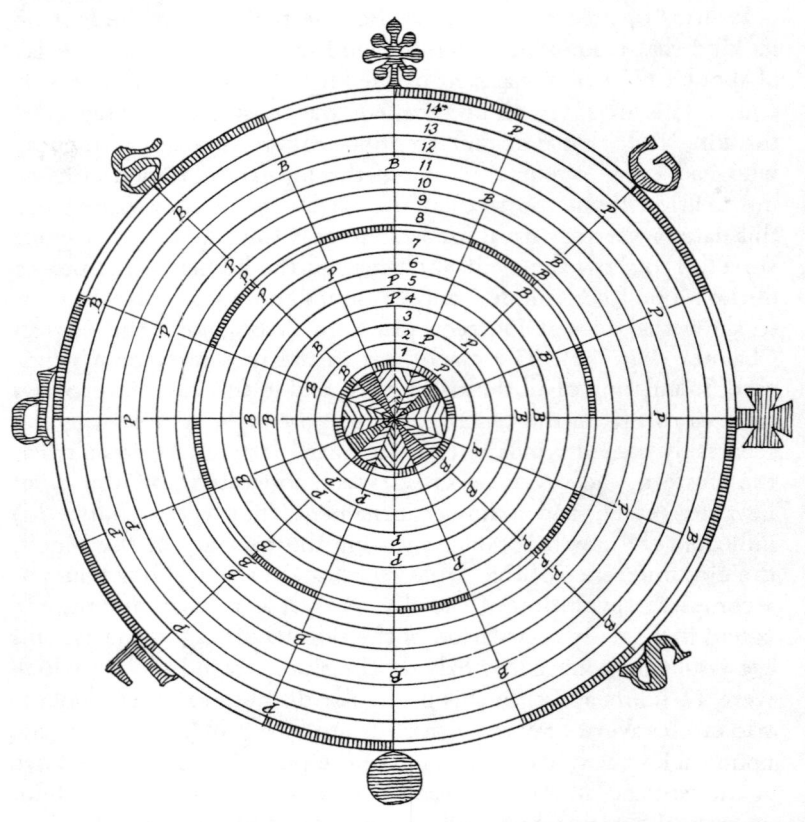

Fig. 16. The Catalan diagram of the Establishment of the Port. 1. Raz de Sein; 2. S. Matthieu; 3. Le Four; 4. Ile de Batz; 5. Sept Iles; 6. Guernsey; 7. Raz Blanchard; 8. Portland; 9. Wight; 10. Beachy Head; 11. Winchelsea; 12. Romney; 13. Sandwich; 14. Mouth of the Seine.

To the modern reader, of course, the Catalan tide diagram appears as the earliest known statement regarding the 'establishment of the port', the time lapse between the theoretical and the actual occurrence of high tide. But there was as yet no accepted theory of the tides, and sailors at least had no reason to suppose that on the days of conjunction and opposition the tide 'ought' to occur at noon. Roger Bacon had, indeed, conceived of some such theory in the middle of the thirteenth century, and had explained it in his *Opus Majus*, written in 1264 at the request of the Pope. He had read (he said) the account of the tides given by the Arab Albumazar in his *Introduction to Astronomy*, but this author did not say what was the cause of the tides, although he had pointed out a fact that was generally agreed, namely that when the Moon produced an effect in one quarter it produced the same effect by reflection in the opposite quarter. This explained why there were two tides a day. Roger himself considered that the Moon's rays drew up vapours from the ocean depths and this caused the ocean to swell. The most powerful rays were those that fell direct from an overhead Moon, but the rays near to the direct ones also had drawing power. Hence the tide began to rise when the Moon was south-east, and continued to rise until it was south-west, after which the rays became too oblique and so the sea surface sank back again until the tide due to reflection began. This theory would give high water south-west–north-east, as on the coasts from Spain to Brittany, and there seems little doubt that this was accepted as the 'normal' time of high tide at the beginning and in the middle of the Moon's cycle. Roger Bacon was, as is well known, anxious to find a mathematical or geometrical interpretation of phenomena rather than to rely upon explanations based on the occult influences of the planets. But Bacon was before his time, and there is no reason to suppose that he was read in Majorca, nor did it occur to sailors to do other than take the tides as they found them, directed by the finger of Providence. Only the Mediterranean traveller grumbled that whereas at home if a ship wanted to anchor at night all one had to do was to shelter behind a rock against the wind, once she got out into the Atlantic Ocean the morning might show her anchorage lying a mile or more in from the water's edge.

Nevertheless, the tides had their uses, as the following story suggests. In 1324 an English fleet of *nefs* set out to carry soldiers to a minor war in Gascony, then under the English crown. Leaving Plymouth the ships were almost in sight of the Breton coast when a southerly wind drove them back across the Channel, and they took shelter in Falmouth bay, anchoring in the creek that runs up to Penryn. It was afternoon, and the men had gone ashore when news came that a fleet of eleven galleys flying the Genoese flag had been sighted entering the bay. The soldiers came tumbling aboard, some shouting for the assault, others crying, 'Peace! Peace!' And the Genoese must have been equally taken aback at discovering a warlike navy in a Cornish harbour. For as it turned out they were peaceful traders laden with wool for Flanders. And there was actually a member of the famous Florentine House of Bardi—moneylenders to the English king—on board. So they stood well away. But, says the Englishman who writes the official despatch, whilst he and some of his fellows put on a bold front, 'we could not approach them or do them any damage, for by then our ships were aground as the tide went out'. The Italians sent their admiral across in a small boat to explain who they were, but when darkness fell they prudently slipped away leaving him behind.

It was on the Atlantic Ocean that Portuguese pilots slowly and painfully learned new methods of navigation during the fifteenth century. But meanwhile the Mediterranean pilots continued to follow their customary rules, only adding to the ancient *Compasso da Navigare* appropriate sections covering the French coast and the Narrow Seas in which tides, depths and grounds took a prominent place. But they did not neglect distances, whereas in the English rutter which has been so freely quoted there is not so much as the mention of a kenning. In so far as they measured their courses it must have been by time sailed, for there is just one passage which says: 'Then must ye go south a glass or two because of the rock.' At a guess, the sand-glass or 'dyoll', as it was at one time called, was turned hourly for keeping course, for that is the way of reckoning met with at a later date. The 'dyoll' or 'horloge de mer' appears in the inventories of English ships from 1295 onwards, but of course, apart from the reckoning, it was also used for setting and keeping the watch, when it was turned every two hours (a 'half-watch glass'). There is all too much room, however, in writing of medieval navigation for doubt and speculation. Yet when a brief glimpse of the seaman at work can be obtained through the eyes of some Renaissance traveller it appears that he was proceeding much as has been supposed, for the majority of sailors were still medieval in outlook and practice in the fifteenth century, lagging behind the new learning and new curiosities of the leisured classes.

Such a traveller was the Spanish gentleman Don Pedro Niño who undertook a voyage in about 1406 from the Mediterranean to the English Channel and whose adventures and observations were recorded at a later date in the chronicle called *El Vitorial*. Weather conditions, as was so often the case, compelled him to move his fleet along the Barbary coast, but fortunately 'the captain had good pilots who knew that coast', and when the ships lay at anchor during the night they 'visited all the neighbouring coves, creeks and anchorages'. So we see how their knowledge was built up. The usual conferences of all the experienced sailors were called daily by the captain when the course to be followed had to be decided upon. And among the company was a certain Juan Bueno who all his life had been 'going about in carracks, sailing-ships and galleys'. So his advice had great weight in determining the right moment for crossing into Spain. When that day came 'the sailors all made ready. They prepared their sea-compasses, animated by the lodestone ['*concertaron las brujolas cebadas con la piedra yman*', says the original]; they opened their charts and began to prick and measure with the dividers, for the course was long and the weather adverse. They set the hour-glass and entrusted it to a watchful man.'

This was as a landsman saw it, but it does appear a sound preparation for a voyage to be made by the Rule of Marteloio, based on distances and bearings, each leg of the traverse being timed. The ships were also prepared: 'Sails were hoisted in the galleys, tillers were fixed, and the oars shipped. They sailed all day and all night, and at dawn sighted Spain, the wind having veered to the south.' All went well and soon they were crossing the Bay of Biscay. Here sails and oars were used alternately, according to the wind, and they stood well out in the open sea, as of course the sailing directions enjoined. The narrator says that the sailors 'feared being cast on the Maranzim' and in fact the unbroken sandy shore of the Landes with shallows extending far out to sea is one that may well be dreaded. 'Then they made a calculation, according to the length of time they had been on that course, that they must be beyond all those perils. They did not know the latitude, but heaved the lead, and found sand, although it was a rocky bottom, and concluded that they were near land.' In fact they found the continental shelf edge and made for Brest. From Brest they sailed round Pointe St. Matthieu, apparently without difficulty, and reached St. Malo. But in leaving this port they had to cross the race. Having the wind astern they went out on the ebb, but were still only half-way across when the tide turned. The water 'boiled up in whirlpools' and they had to struggle with their oars, in great danger, for six hours until the ebb set in again. 'The Western Sea', laments the narrator, 'is not like the Mediterranean Sea, which has neither ebb nor flow nor great currents, if one excepts the Faro current [i.e. in the Strait of Messina] which is very dangerous, and where many ships have perished.' The Ponente, the Western Sea, was, he considered, 'most evil, especially for galleys'. And indeed, as they had rounded Pointe St. Matthieu the alarmed sailors had called variously upon St. Mary of Finisterre, St. Vincent of the Cape, and St. James of Compostella, according to their particular devotion, begging for protection and promising gifts to the shrine. Just so had the ancient Greek sailors called upon the gods and goddesses whose temples had stood on the promontories, re-placed now by the chapels and images of the saints.

Pedro Niño experienced no serious mishap, but another traveller, the Magnifico Messer Piero Quirinos, a Venetian, was not so fortunate. Mankind, as he said himself, was bound to the spokes of blind Fortune's wheel, and as she turned it he was now lifted high to success, now plunged low to misery and despair. Bound for the Flanders Channel, his ship set out from Candia (Crete) in April 1436, and because of contrary winds took over a month to reach Cadiz. Approaching this port an incompetent pilot drove her on to a hidden rock, and there was a long delay for necessary repairs. Meanwhile Messer Piero learned that his country was at war with Genoa, and fearing he might encounter a Genoese ship, took an extra score or so of soldiers aboard, so that his own ship was greatly overloaded. The next misfortune occurred as they rounded Cape St. Vincent, when a contrary wind drove them nearly to the Canaries. In fact by the time they made Cape Finisterre seven months had passed, and it was

November. However, some of the company had landed in Galicia to visit the shrine of St. James of Compostella, and for a time Fortune appeared to smile. They made two hundred miles across the Bay, and taking soundings came abreast of Ushant. But when they should have turned north-east to enter the Channel, a strong south-easterly wind blew and they found themselves outside the Scillies. Running before the wind, which did not abate, they arrived off the south-west of Ireland and encountered two sea-going curraghs, but as they knew not a word of Erse they could get no help from those aboard as to their whereabouts. The storm increased, a mast went overboard, and the rudder was torn away, so that they had to rig a jury rudder. Taking soundings again, they found 80 fathoms, and quickly 120. An attempt to anchor by joining three cables was a failure, and the pilots declared (apparently on the evidence of the soundings) that they were 500 miles from the nearest land. In fact one of the two narratives of the voyage (which were printed by Ramusio in his great collection) says they were 700 miles west of Ireland. To cut a long story short, the wind changed and blew furiously from the south-west, and after an interval of terror and helplessness, with provisions all consumed, one of the company saw a shadow on the horizon before the prow, which they dared not believe was land. However, when dawn broke they could see high snow-covered mountains and cast themselves ashore on Rusten Island in north Norway.

A third voyage, the most illuminating of the three, took place in the opposite direction. An intelligent German monk, Brother Felix Faber, was going on a pilgrim ship to the Holy Land and Egypt. He observed very carefully all that he saw, and related it in Latin to his brethren on his return in 1483. Having little to occupy him on board he devoted several chapters in his narrative to the handling of the ship. 'There was a powerful officer aboard ship', he wrote, 'who was called the pirate, or as we Germans would say, the pilot. This pirate knew the shortest and safest routes across the sea, and consequently the voyage was made as he directed or advised. If, however, they came into any area which he did not himself know he made for the nearest port and handed over his office to someone who knew the paths of the sea there.' In fact he handed over to a petty or local pilot as maritime law compelled him to do. 'Besides the pilot', Felix continued, 'there were other learned men, astrologers and watchers of omens [auruspices] who considered the signs of the stars and sky, judged the winds, and gave directions to the pilot himself. And they were all of them expert in the art of judging from the sky whether the weather would be stormy or tranquil, taking into account besides such signs as the colour of the sea, the movements of dolphins and of fish, the smoke from the fire, and the scintillations when the oars were dipped into the water. At night they knew the time by an inspection of the stars.' The presence of learned men on board is intriguing, especially as they were telling the time by the hour angle of the stars. But so could common sailors by that date (at least roughly), and it is impossible to say whether such men had usually been carried on important ships with important passengers (as this ship was) in the truly medieval period, and particularly when the 'new navigation' began. It is at least very probable that they were.

Faber now goes on: 'And they have as compass a Stella Maris near the mast and a second one on the topmost deck of the poop. And beside it all night long a lantern burns, and they never take their eyes off it, and there is always a man watching the star [i.e. the compass rose], and he sings out a sweet tune, telling that all goes safely, and with the same chant directs the man at the tiller how to turn the rudder. Nor does the helmsman dare to move the tiller in the slightest degree except at the orders of the one who watches the Stella Maris [the magnetic compass] from which he sees whether the ship ought to go straight on, or curve or turn sideways. And they have other instruments with which to judge the course of the stars, the direction of the wind, and the path of the sea.' This description of the practice of the mariner's compass is a compelling one, and the procedure was no doubt traditional: it fits in exactly with all that the documents suggest about Mediterranean navigation. But the 'instruments with which to judge the course of the stars' may have only been in use because the astronomers were there.

Faber's next section refers to the maritime chart. 'They have a chart on which is a scale of inches showing length and breadth, on which thousands of lines are drawn across the sea and on which regions are marked by dots and numbers of miles. [A description of the rhumb lines more impressionistic than correct.] Over this chart they hang, and can see where they are even when the stars are hidden. They find this out by drawing circles on the chart from line to line and from point to point with wonderful industry. [Actually the pilot was probably employed with ruler and dividers or drawing compass in plotting his course.] They have many other instruments', continues Felix, 'with which they study the way across the sea. And they sit in these conferences every day. The padrone or governor of the ship [i.e. the captain] does not interfere in matters of navigation, nor does he know the art, but he orders the ship to go this way or that.'

The vessel was a medium-sized galley with oars and sails, and the monk explained that the rear part was called the poop, upon which was built a structure called the castle. The rudder hung from behind the poop into the sea, and the helmsman, tiller in hand, sat above it in a latticed cabin. The castle had three decks. On the uppermost was the Stella Maris and the man (probably the mate) who gave the orders to the helmsman according to the compass; there too were the star and wind watchers, and those who found out the routes across the sea (in fact all the navigating officers). Below was the cabin of the lord captain, and of his table companions, and below again was the captain's treasury, and here, too, noble lady passengers slept at night. The only light entering this chamber (if so it can be called) was through holes pierced in the deck above, and it must have been a fearsome place, since no lights were allowed at night except the compass lantern. Faber says that the steersman's cabin was also on the topmost deck, and this may have been the case on a galley although it was on the deck below in a caravel. The generality of pilgrims and other passengers slept in the hold, reached by an open hole and a gangway near the main mast. Conditions do not appear to have been nearly so comfortable as on the Chinese ships on which Marco Polo travelled, which had numbers of separate cabins, as had large Arab ships.

Telling the time by the stars (referred to above) had been systematized ever since Ramon Lull's days. The observer had to know the midnight position of the Guards of the Lesser Bear for each month (or, better, for each fortnight) of the year, this

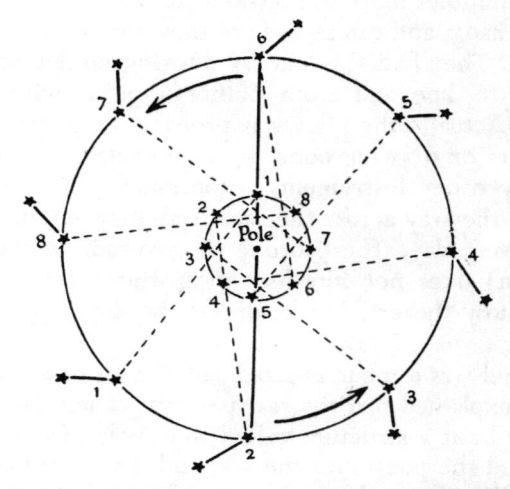

Fig. 17. The Guards and the Pole Star. When the Guards are upright the Pole Star gives the height of the Pole within half a degree.

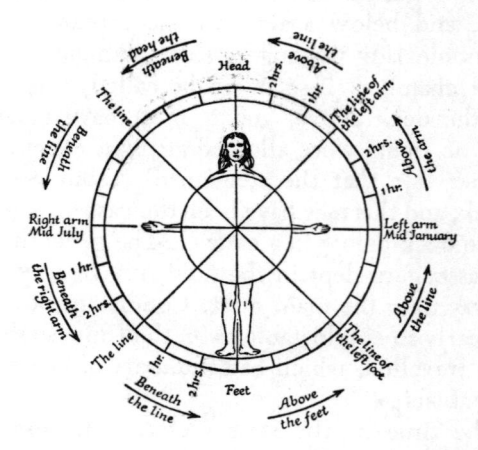

Fig. 18. The sky clock. The midnight position of the Guards is memorized for each fortnight of the year and a comparison of their observed positions with this gives the time in hours before or after midnight.

position shifting about an hour every two weeks (Figs. 17, 18). Supposing the Guards are approximately on the meridian beyond the Pole Star at midnight, then by the time they are due west of

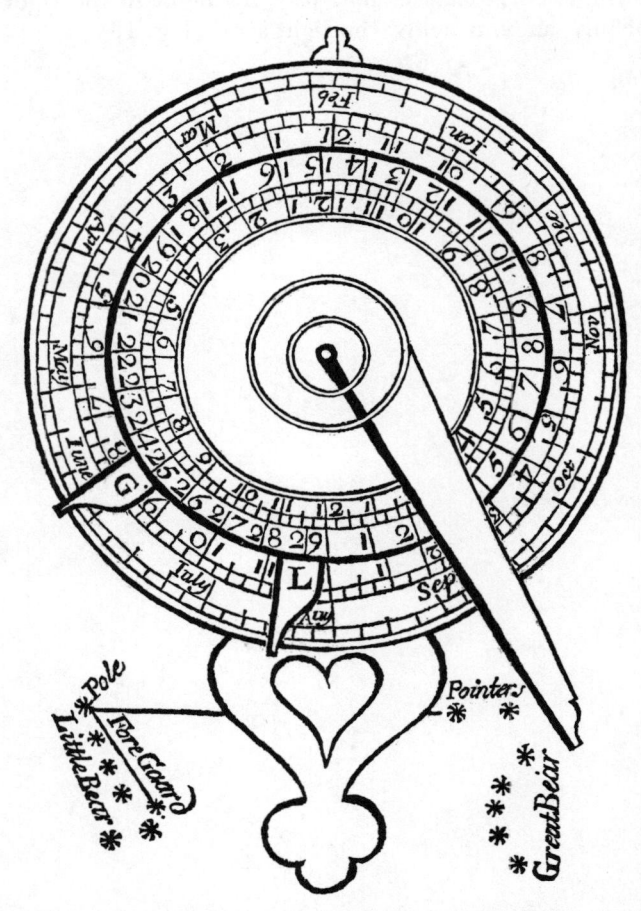

Fig. 19. A nocturnal with scales to be used with the Great and Lesser Bears.

the star it will be 6 p.m.; and at 6 a.m. they are due east. An instrument—the nocturnal—was later designed for observing these positions accurately (Fig. 19), but sailors imagined a human figure up in the sky with the Pole Star in his breast. His head was 'above' it (i.e. north), his feet 'below', his arms to right and left and the Guards were described as they stood in relation to his limbs. For example 'mid-July, midnight in the right arm. End of July, an hour below the right arm' (Fig. 18).

PART FOUR

INSTRUMENTS AND TABLES

VII

The Portuguese Pioneers

ALTHOUGH sailors were quite unaware of it, the recovery at the beginning of the fifteenth century of Ptolemy's *Geography* and maps by the western world was as vital an incident for them as it was for the cosmographers and cartographers. The main part of Ptolemy's text is a list of latitudes and longitudes covering the world as he knew it. He explains, too, how to construct a network of lines of latitude and longitude for a map on a conical projection, supposing a degree of the arc of the meridian, and of the equator, to measure 500 stadia. On such a projection and network his world map was drawn—whether by himself or by his disciple Agathadaimon makes no matter. The Greek manuscript of his work, brought from Constantinople, was translated into Latin in 1409, copies of the maps were made, and the learned world of Renaissance Italy (with Germany not far behind) was soon convinced that the only reasonable way to fix position on the terrestrial globe was by such a system of co-ordinates, analogous to those used to map the heavens.

The idea, of course, was not new to astronomers, who were building up a growing list of latitudes and longitudes to be used with their Ephemerides. But only Roger Bacon, a hundred and fifty years earlier, had tried to use them in drawing a map. And in fact the astronomers did not always make and use such lists in such a way that the layman could recognize them as longitude tables. Their purpose was so that each observer could 'rectify' the observations and tables made, say, at Toledo, for use in his own city, say London. All that was necessary was to know the 'hour-angle' or time difference between the two places, which is four minutes for each degree of longitude. Each heavenly event occurred proportionally earlier by local time if the second observer lived west of the place where the Ephemerides were drawn up, and proportionally later if he lived to the east. But in actual fact it is hardly correct to speak of the observer, for the

Ephemerides were drawn up mainly by calculation of the motions of the Sun and planets and were chiefly used for prognostication of good and evil hours. Ptolemy had pointed out, and Hipparchus before him, that by noting at two places the time at which a very obvious heavenly event, like an eclipse of the Moon, was seen, the local time difference, and therefore the longitude difference between two places could be found. But since there must be a competent observer at either place furnished with Ephemerides, and each must have a precise means of measuring his local time, it is not surprising that Ptolemy could cite only one very poor example of such a calculation, and for his own part based all his longitudes on the measurement of distances. If the latitudes of two places A and B are known, and the distance between them,

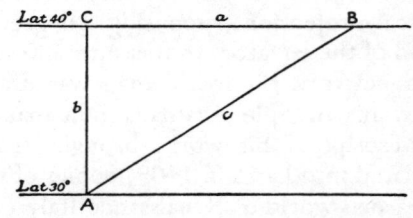

Fig. 20. Pythagoras' theorem $(a^2 + b^2 = c^2)$ was used for finding difference of longitude, a, supposing difference of latitude, b, and distance, c, known.

then, by using Pythagoras' theorem, the east–west distance can be found (Fig. 20). And since the size and shape of the globe were known to him, Ptolemy could translate the east–west distance into longitude difference for the particular latitude of B. An error is introduced by treating what is in fact a spherical triangle as a plane Euclidean right-angled triangle. But, as the initial errors (save in very few cases) of latitude and distance must have been considerable, it is of no importance to decide whether Ptolemy corrected his figures for this error. Sailors, however, even when they began to use latitude, continued for a century and more to be satisfied with east-westing directly measured. It was Pope Alexander VI who, by decreeing a meridian line of partition between Portugal and Spain in respect of their discoveries in the western hemisphere, made longitude a live issue once the rivals had met and disputed the Spice Islands in the eastern hemisphere. Long before that happened, however, another difficulty had been introduced by this putting of geography into a mathematical framework. The equator is a precisely defined line midway between two poles that are also precisely defined; that is to say their positions can be actually identified on the ground. Consequently the circles of latitude can be numbered in degrees from 0 to 90 northwards and southwards from equator to poles. But the meridians, or circles of longitude, each running between pole and pole, are all alike and indistinguishable. Ptolemy (arguing against the opinion of his predecessor Marinus of Tyre) had been able to convince himself that the known world occupied precisely and neatly half the circuit of the globe from west to east, so he numbered off his meridians from 0° to 180° beginning with the farthest land reported in the west, which happened to be the Canary Islands, then known as the Fortunate Islands. The Arabs, too, reckoned that they had only 180° to deal with, counting backwards from the place where the Sun first rose at the Creation and the Earthly Paradise stood. But their tables of longitude (and the Christian ones derived from them) in some cases ran to what they called the 'farthest west', and in others only to the 'farthest inhabited west', which differed by 17° 30'.

Longitude, in fact, has suffered from great confusion for lack of a fixed prime meridian. Fortunately up in the sky there are two fixed points on the equinoctial line (or heavenly equator) from which east–west measurements can be made. These are the points at which the Sun's path—the ecliptic—crosses the equinoctial at the spring and autumn equinoxes, and they can be identified by the neighbouring fixed stars. It is true that these equinoctial points shift very slowly through the centuries, so that the fixed stars (e.g. the Pole Star) gradually seem to us to change position, but it is the astronomer's business to allow for this in his tables. Unfortunately for the layman, the early astronomers were (no doubt very naturally) more concerned to follow the Sun round the ecliptic than to consider the division of the equinoctial, from which it 'declined' alternately north and south. Consequently, whereas distance north or south of the terrestrial equator is called latitude (for it measured the breadth of the known world), the corresponding distance of heavenly bodies from the equinoctial is called declination. And whereas east–west distance round the equator is called longitude (for it measured the length of the known world in Ptolemy's day), the corresponding sky-term is right ascension. The fixed stars have a certain declination and a certain right ascension as places on the Earth have a certain latitude and longitude, while differences of both right ascension and longitude can be treated as 'hour angles' or differences of time (1 hour = 15°). On the other hand the position of the Sun and the planets were defined in relation to the ecliptic and the constellations or Signs through which this circle ran. Each Sign governed 30°, so that the Sun, which is always on the ecliptic, was completely defined by saying its position was $9\frac{1}{2}°$ in Virgo, or 27° in Capricorn: but the planets had to have their distances away from the ecliptic defined as well as their degree in the Signs. And this distance was their latitude. A medieval speaker would just as naturally date an event by saying 'The Sun was in the ninth degree of Cancer' as a modern man would say 'It was on the twenty-first of June'. An astronomer, of course, could easily calculate the Sun's declination from the equinoctial from its position on the ecliptic, provided he knew the angle between the two circles. But observers differed somewhat as to the exact size of this angle, partly because it too changes very slowly through the ages. So tables of solar declination do not precisely agree. Moreover the Sun takes rather over a day to get through each degree of the zodiac, and the calendar has to cover four years before the figures repeat themselves, while the cycle of the dates of new and full Moons is only completed every nineteen years. Altogether it was a very serious matter to introduce astronomy and the calendar to the sailor, although, as has earlier been suggested, it became necessary by the fifteenth century to do so, for the known world was expanding fast and voyages growing correspondingly longer.

The notion of the Earth as a sphere had become widely familiar among educated people by this time. Most university students read Sacrobosco's elementary textbook on the subject, while many men and women of culture read and enjoyed Sir John Mandeville's *Travels*. Here they found familiar mention of Antipodes, and learned that in this knight's opinion, if only men and ships could be found, the circumnavigation of the globe could be carried out with no greater difficulties or dangers than were encountered on any other long voyage. To those who had travelled in the East, or had simply read Marco Polo's entrancing book, Japan (Cipan-gu) was already known by hearsay, although placed 1500 miles beyond China, while to the west the Azores Islands and the Madeira group had mysteriously appeared on the charts used for sailing to England and Flanders, although there is no record that they were officially 'discovered'. The Mediterranean sailors who saw and charted them, presumably when they were forced to run west before the wind during bad weather, took them to be St. Brendan's Isle, or the Fortunate Islands, or perhaps Brazil Island which, so they had heard, lay out to the west, and thought no more about them. And the warning Pillars, which according to tradition Hercules had set up advising mariners not to trespass into the outer ocean, were being painted by the chart-makers farther and farther out to sea, associated now with the name and dread of Satan. Towards the far north, again, there were supposed to be many great islands that the British king Arthur had conquered and colonized, as could be read in the *Gesta Arthuri*, while in 1364 an English friar had written an account of this quarter of the world based on his travels westward from Norway. He appears (for the book is lost) to have made visits to Iceland and Greenland, and possibly to the Vikings' Vineland in North America, and at the very Pole, he said, stood a huge black magnetic rock, towards which venturesome ships would be hurried by four indrawing seas rushing between four great islands. And they could never return against the current but would be lost.

Southwards, however, the known world had somewhat contracted since the days when Ptolemy had drawn Africa stretching sixteen degrees or more beyond the equator, and even then not reaching the Southern Sea. And the Terra Australis shown on Crates' symmetrical globe had been almost forgotten. Most cartographers swept a curving line round from somewhere south of Morocco to somewhere south of Abyssinia, which made the circumnavigation of Africa look little more formidable than a voyage from one end of the Mediterranean to the other. There was a story that two Italian brothers Vivaldi had made it from the west, and a relative went to look for the reputed survivor in Abyssinia. On the great Catalan Atlas of 1375, too, one James Ferrar was painted in his sailing galley setting out southwards beyond the Pillars to seek the River of Gold which lay somewhere on the way round Africa. Actually the Moors were already fishing and trading along the West African coast as far as the Gulf of Guinea (where on one occasion a Spanish friar accompanied them), while in East Africa Arabs were now settled as far south as Sofala. But this did not appear on the maps, although one or two carried a strange, nameless, featureless South African peninsula.

The maritime chart, however, knew nothing of all this. It stopped abruptly at Cape Nun, while showing not far off-shore a correctly charted and named island group, the Canaries, long ago described by King Juba of Mauretania. They had been neglected during the middle ages, although the Moslem Wanderers, or Maghurins of Lisbon, were said to have visited them, but for the most part the Arabs left the Sea of Darkness severely alone. A Genoese built a castle there in 1270, but actually the islands only became part of the European world when in 1402 they were occupied by a Norman knight, the Sieur de Béthencourt, and were subsequently assigned by the Pope to the Crown of Castile. This was a sore point with the Portuguese, who having expelled the Moors from their country were contemplating an attack on them in Africa. It was, in fact, at the successful capture of Ceuta in 1412 that Prince Henry the Navigator, third son of an English Queen of Portugal, and grandson of John of Gaunt, first won renown when barely of age. Subsequently he became a member and head of the religious Order of Christ, and vowed his life to Africa and the solution of its problems. Among these there was the River of Gold to be discovered, and the fabulously wealthy King of Ghana, whom the Catalan maps depicted as enthroned in his kingdom beyond the Sahara. There must, too, be heathen converts to be made, and there was even the possibility of finding a way from the west to the Christian kingdom of Prester John, cut off behind the Moslem kingdoms of the middle east.

That Henry ever had in mind the remoter purpose of opening a sea-route to India is very doubtful; certainly his biographer says nothing of it. Indeed, the point he emphasizes to explain the prince's life is that his horoscope showed him destined for daring deeds and discoveries, a reminder that astronomer-astrologers were an accepted part of the entourage of an Iberian prince. And the long tradition of patronage of learned Jews still held good in Castile and Aragon. When Prince Henry set about the maritime exploration of Africa, he collected around him pilots and sea-captains, maritime charts, maps and books. But all these were not enough, and in 1420 he sent for Master James of Majorca. It is claimed that Master James was none other than Jafuda Cresques, the instrument-maker, compass-maker, and chart-maker, who forty years earlier had, like his father, served the king of Aragon. Certainly these were just the skills that could not be found in Portugal yet were essential to the prince's purpose. And the elderly Jew when he arrived could set them all on foot. He would have found the leading Portuguese pilots already trained in the Mediterranean technique of navigation, for in 1317 the king of Portugal had appointed a Genoese sea-captain, Manuel Pessagno, as hereditary admiral of his fleet, and the Italian had brought in a score of his fellow-countrymen who were skilled masters and pilots, to work under him. Native Portuguese, too, who traded to Galway and Flanders, were familiar with the 'lead and line' methods of Breton and British sailors. Such skills, however, whether Mediterranean or Atlantic, depended upon centuries of past experience, observations and records, and were insufficient for new navigations in new waters. Indeed the mariners timidly feeling their way along the African coast for long declared it was impossible, even suicidal, to attempt to round Cape Bojador, little more than 200 miles beyond Cape Nun. And one thing that must have alarmed them was the steady set of wind and current in those latitudes away from home.

But farther north the case was different. Between the latitudes of Lisbon and the Canary Islands the prevailing summer winds are from north-east to north-north-east and a ship sailing before the wind will necessarily encounter Porto Santo and Madeira. These islands were in fact actually colonized by 1420, for here there is no trouble about the return journey, for the winds become light and variable as the Sun declines to the south, and during the winter southerly and south-westerly winds are frequent. Moreover, within ten or a dozen years of the settlement of Madeira, the Azores, too, were visited and then colonized—a group of islands lying a third of the way across the Atlantic Ocean. And only after these events, that is to say in 1433, did one of Henry's captains, urged by his master to take no notice of ignorant mariners, sail round Cape Bojador on a second attempt. Portuguese ships were now sailing some 700 miles west and 700 miles south, and often farther afield still to catch the wind. Fortunately a new sailing technique to meet their needs had been devised, perhaps by Master James, certainly by the advice of astronomers. It sounds at first hearing very simple. Dead reckoning was to be checked by instrumental observation of *altura*—the height of Sun or star—for the word latitude was not mentioned, and ports were sought also by *altura*.

To the ordinary man of the day the astronomer was, quite simply, a magician, his instruments the means of magic practice. And his elaborate and costly astrolabe (Plate IX), by which he read the skies, was an instrument that could not possibly be put into the hands of a pilot who could do no more than master the Rule of Three. The main device of the astrolabe was, in fact, beyond the comprehension of any non-mathematician. A stereographic projection of the heavens with the star positions engraved at points on a pierced brass plate turned above a stereographic projection of the observer's view of the sky, as defined by a network of lines of equal altitude, and lines of equal bearing. Round the margin of the instrument were set the divisions of time and of the seasons which depend upon the apparent rotation of the heavens and the seeming yearly motion of the Sun. On the back of the astrolabe, however, there was an alidade or sight-rule turning about the centre, by means of which, when the instrument was suspended from its thumb-ring, the height of the Sun (or of a star) could be observed by means of two pin-holes. The opposite points of the alidade turned on two scales from 0° (the horizontal position) to 90° (the vertical); and sweeping every other part away, the instrument-maker made a 'seaman's astrolabe' consisting only of the main circular plate with its swivel suspension-ring, which carried the alidade and was engraved with the scale of degrees (Plate XVIII).

But even this simple instrument appears to have been at first beyond the pilot's capacity to use, for the earliest mention of taking an observation, which is not until 1456–7, refers to the use of a quadrant. Perhaps, indeed, the new methods were first left to astronomers, for a member of the party which took King Alfonso's sister by sea from Lisbon to Pisa remarked that besides the most famous sea-captains they carried master astrologers who were well instructed in travel by star and pole, that is by *altura*. The astronomers' quadrant goes back at least to the thirteenth century, and like the astrolabe could answer questions about the planets, the time and the seasons. It carried, too (as the back of the astrolabe did), the geometrical square, which was one day to be adapted for sailors' use. But again for sea use all its complicated lines and curves and star positions were taken off from the engraved plate, and all that remained were the little squares carrying the pin-holes for sighting attached to one straight edge, and the plumb-line, a fine silk thread and weight, which fell from the right angle across the scale of 90° along the curved edge, and so marked the angle of elevation of the sights when a star was observed (Plate XIX). The star was of course the Pole Star, for our informant Diogo Gomes, a young gentleman in Prince Henry's service who was sailing to Guinea, says: 'I had a quadrant when I went to those parts. And I marked on the scale of the quadrant the altitude of the Arctic Pole. And I found it better than the chart. It is true that the sailing course can be seen on the chart, but once you get wrong you do not recover your true position.' There were, in fact, cumulative errors if the Mediterranean technique of 'direction and distance' was relied upon for long voyages. Prince Henry's captains, and those pilots who were not too conservative to alter their methods, marked the star altitudes of successive capes and river mouths and islands on their quadrants, and the astronomers at home built up a table of coastal latitudes which by 1473 had reached the equator. But there is evidence that at first, as pilots did not know how to use a scale of 'degrees', and had never thought in terms of 'latitude', quadrants were marked with important place names against particular parts of the scale, and the pilot could thus recognize his position by the fall of the plumb-line alone.

By what must have been the merest chance a rather mutilated

set of very early directions was added to the later edition (1563) of a nautical book compiled in 1518 by a scholarly German printer in Lisbon who had collected notices of the early Portuguese discoveries. And here we read the first instructions (as they must surely be) to a sailor for using an observing instrument—the quadrant. And first he is warned (and it must have come as a shock) that the Stella Maris is not 'fixed' and does not stand at the same height all day long over Lisbon. So he must observe it only when the Guards are in a certain position. Fortunately he was accustomed to watch the Guards, for the front Guard, Kochab, was the hour hand of the sky clock, described earlier. 'Sailor,' says the instructor, 'the quadrant is used in this way. For each degree marked on the scale [tronco] you must count 16 leagues and two-thirds, which is two miles, reckoning three miles to a league.' The sailor at first only understood the linear scale or trunk, as marked on his chart, and the fact that the degree is here said to be $16\frac{2}{3}$ leagues, or only 50 miles, is evidence that the instructor came from Spain. For the Portuguese sailors were in later sailing directions taught to count $17\frac{1}{2}$ leagues, of 4 miles each, or 70 miles to a degree—a much more accurate figure. It almost looks as if the Portuguese astronomers accepted Eratosthenes' degree of 700 stades, while the Spaniards and Catalans chose Ptolemy's of 500 stades, both reckoning 10 stades to the mile, but each using the local league.

To return, however, to the instructions. The sailor is told that he must observe the star through both the pin-holes, like a crossbowman taking aim with his bow, and then mark where the lead falls on the scale. First of all the mark must be made for Lisbon, and then he must mark the successive places on the voyage. And every observation is to be made when the guards lie east–west (Fig. 17) with the star. Then, if he wishes to go to a certain place, he should look for its name on the scale of the quadrant—say Cape Verde or Finisterre, and when the lead falls on that degree then he is east–west with that place. Here, then, we see the beginning of the new navigating method—finding and running down the latitude—which became standard practice for centuries. These early, and crude, directions go on to explain also how to sail by the height of the midday Sun, once again by an ingenious simplification which was later to become unnecessary as pilots became more skilled in computation. Briefly the observer was furnished with a table of daily solar altitudes at Lisbon, 'or at Madeira or any other port of departure', for the hour of noon, and the difference of altitude that he found gave directly his difference of north or south distance (using the scale of degrees to leagues) or the difference of latitude as we should say. The astronomers of Lisbon could readily calculate the necessary altitude tables from existing data in their Ephemerides, although the sailor could not be very precise.

In observing the Sun the instrument has to be held until the Sun's rays pass through both holes on the sight rule and cast a spot of light on a surface behind it, and this could be more easily done with the astrolabe hanging from its ring than with the quadrant grasped in the hand. But the difficulties arising from the motion of the ship and from the wind were formidable, and where possible seamen landed to make their observations. Nor is it surprising that instruments were very slow in coming into general use. The Italian Cadamosto, who led an exploratory voyage south in 1454, writes entirely in terms of older methods. After leaving the Canaries the ships sailed for a couple of hundred miles in the open ocean with the trades, and then turned in to the shore and coasted with lead and line, lying up at night. 'At dawn we made sail, always stationing one man aloft and two in the bows of the caravel to watch for breakers which would disclose the presence of shoals.' They did indeed also watch the Pole Star, but not with instruments: 'It appeared about a third of a lance above the horizon', Cadamosto reports, when they were in the mouth of the River Gambia (13° 50'), while on their way home in about 12° Lat. 'the Pole Star appears at the height of a man above the sea.' When in the Gambia, too, he says that on July 2nd, 'we found the night to be 13 hours and the day 11 hours' (figures which should be reversed), a reminder of the way the Greeks defined latitude. Cadamosto and his companion the Genoese Usodimare were not, of course, professional navigators, although they had had previous experience at sea, but as they would take part in the daily consultations with masters and pilots it is unlikely that they would have overlooked any instrumental observations, had these been made. It was when in the Gambia that Cadamosto first reported the sight of the Southern Cross: 'This we took to be the Southern Wain [or Bear], though we did not see the principal star [i.e. the Antarctic Pole Star] for it would not have been possible to sight it, unless we had lost the North Star.' Here he is reflecting the common belief that there was a motionless Pole Star like an axle-tree at each of the points about which the heavens turned. Indeed, the name Axis was actually given to the North Star, while the phrase 'Antarctic Pole' was used by astronomers to signify the whole southern hemisphere which lay under that pole.

Prince Henry the Navigator died in 1460, but the exploration of the Guinea coast was continued, and by 1474 the equator had been reached and the islands of S. Thomé, Principe, and Fernando Po placed on the chart. The conduct of African affairs was then assigned to Prince John, but political troubles, including a war with Castile, prevented any serious attention to problems of navigational technique until 1480. Like his uncle, King John (he succeeded in 1481) had astronomers and mathematicians about him to whom he now gave in particular the task of improving the finding of altura (as it was always called) by the Sun. For in 1481 the castle of St. George was built at Mina in the Gulf of Guinea, where as a contemporary chart notes 'the Star' was not visible, while all further exploration would be beyond the equator. The names of three of the king's advisers have been preserved by the historian John de Barros, writing many years later (1539). They included a Royal physician, Master Rodrigo, a Royal chaplain, Bishop Ortiz, and a learned Jew, José Vizinho, who was the disciple of a famous astronomer, Abraham Zacuto of Salamanca, who ten years or so later himself came to Lisbon. The results of their work are to be found in the oldest surviving navigation manual, the Regimento do Astrolabio e do Quadrante, a copy of which was printed in Lisbon about 1509. That this was not the first edition of the manual any printer would detect, and an earlier printed edition of about 1495 has been suggested. But handwritten copies would have been prepared at once for the selected and trained pilots, and for the captains of the ships commissioned to make important voyages. The fact that the manual contains a list of latitudes only going as far as the equator puts it back to 1480–81, and so it will be examined as exemplifying the practices taught at that date, which in fact remained the basic techniques for the next century or more.

The essential contents of the manual were the Regiment of the North (i.e. the rules for observing the North Star), the Rule for Raising the Pole (replacing the Rule of the Marteloio), the Rule of the Sun, a list of *alturas* from the equator northwards, and the calendar for the year commencing on March 1st, which gave the Sun's position in the Signs, and its declination day by day.

In the earliest rules, it will be remembered, the Star was always observed at one particular position of the Guards, that is to say when they were 'east–west', and at the end of this explanation the instructor added that with 'Guards in the head' the

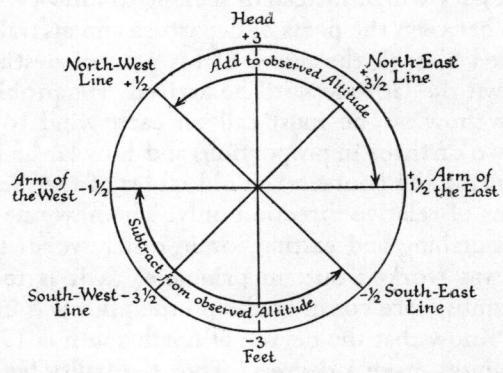

Fig. 21. The Regiment of the North Star. The eight positions are those of the Guards and the figures show the number of degrees to be added to or subtracted from the observed height of the Pole Star for each position.

reading would be 3° too low, while with 'Guards in the feet' it was 3° too high. The 'sky-man', used for memorizing the star-clock, had in fact been adapted for this new purpose. But in addition to his head and feet (north and south), his right and left arms and four lines midway between his limbs were added to give altogether eight points of reference. At each of these points according to the new rule so many degrees were to be added or subtracted from the observation of the Star to give the true elevation of the pole (Fig. 21). Since, however, the 'sky man' is looking down at the observer with his head towards the north horizon, his right hand was opposite the latter's left, and vice versa, which was extremely confusing. One advance made in the manual was to substitute the terms 'arm of the east' and 'arm of the west' for left and right arms, but it is worth notice that Christopher Columbus used the older terms, while his various mistakes in recording observations suggest that he was only familiar with the older sets of rules. As an example of the clearer rules the third may be quoted: 'Item, when the Guards are in the line below the arm of the east, the Star is above the pole half a degree. The half-degree must be subtracted and the remainder shows the number of degrees by which the observer is separated from the equinoctial line.'

A new rule for 'raising a degree' or resolving the course had become necessary when, instead of seeking to follow a direct line and distance between the ports of departure and arrival, the pilot was instructed to seek the *altura* of his port of destination and then run down the latitude until he arrived. His problem, then, was to know how far he must sail on each wind to raise one degree (or two or three in proportion) and how far he had easted or wested during the course. The old tables of the *Marteloio* had been in terms of relative direction only. The new one had to be in terms of northing and easting, or in other words the course made good was worked out, in principle, as it is today. Only the eight rhumbs were considered, and the pilot was informed at the outset: 'Know that the degree of north-south is $17\frac{1}{2}$ leagues, and sixty minutes make a degree.' Thus the faulty figure of $16\frac{2}{3}$ leagues had been abandoned. The table was given in words but it can be set out as follows:

Wind	Distance (leagues)	East–west (leagues)
First quarter	$17\frac{5}{6}$	$3\frac{1}{2}$
Second quarter	$19\frac{3}{6}$	$7\frac{2}{3}$
Third quarter	$21\frac{1}{3}$	$11\frac{5}{6}$
Fourth quarter	$24\frac{3}{4}$	$17\frac{1}{2}$
Fifth quarter	$31\frac{1}{4}$	$26\frac{1}{6}$
Sixth quarter	$46\frac{1}{2}$	$42\frac{1}{2}$
Seventh quarter	$87\frac{1}{6}$	85

Later editors of the manual found it necessary to add that when navigating on the eighth rhumb or quarter, one did not raise the degree, but kept always at the same *altura* as the port of departure—a reminder that learning was by rote and not by reason.

And bitter experience led to the added warning that despite these rules and figures great errors could result owing to currents. Unknown currents, in fact, bore the chief of the blame for faulty reckoning, although the new method of sailing, based on astronomical orientation, necessarily made pilots aware that their compass-needle 'north-easted' and 'north-wested' at times, they knew not why. The early manuals, however, do not concern themselves with the magnetic compass.

It is the actual figures for the daily declination of the Sun in this first manual which show that they were not derived, as was once thought, from the Ephemerides or from the *Liber Directionum* of the famous Regiomontanus, printed in 1474 and 1476 respectively. The German astronomer took the inclination of the ecliptic to be 23° 30', but José Vizinho's master, Zacuto, took it as 23° 33', and this was the figure adopted, although Zacuto's *Almanach Perpetuum* (with radix 1473) was not printed until considerably later. To teach the rule of the Sun to a novice must have been a matter of great difficulty for an astronomer. The would-be pilot must learn how to 'enter' the calendar and pick out the figures, he must memorize the dates of the equinoxes and know which are the northern and which are the southern Signs. For according as the Sun's shadow falls north or south, and according as it is in the same or the opposite hemisphere to himself, the figures are to be manipulated differently. In fact he has eight rules to learn, apart from the special case when his ship is on the 'line' precisely at one of the equinoxes. He must begin, he is told, by taking the height of the Sun by his astrolabe or quadrant (Plates XVIII, XIX), 'and this must be at midday when the Sun is at its greatest elevation'. He is to keep the altitude found (he probably set it on a slate as Chaucer did) and enter the table for the month and day. 'Take out the declination, and if the Sun is in a northern Sign, and if the shadow is falling to your north, then subtract the altitude that you found from ninety, and add the declination. The sum will be the number of degrees you are north of the equinoctial.' After working two examples the instructor adds 'and if you find an altitude of 90° [when the subtraction and addition would appear to break down] know that you are distant from the line as many degrees as the Sun has of declination, no more and no less'. Next he takes the case where, in the same hemisphere, the shadow falls south, and so on through the rules, posing the special case when on December 12th the altitude of the Sun is found to be $66\frac{1}{2}°$. 'Add this [he says] to the declination, $23\frac{1}{2}°$ and you get 90°, which subtracted from 90° leaves nothing. And so you are right beneath the equinoctial line'. Here there is a discrepancy, since the table gives the declination as 23° 33' but the odd three minutes are ignored.

It is worth noting that in the list of latitudes which the manual provides the positions are, with few exceptions, correct to within half a degree—often to within ten minutes. Later manuals have a more extensive and more carefully thought out list, besides such improvements as a four-year table of declinations, and a table of the tidal retardations during the lunar cycle. The first manual is shown in use in a letter which Master John of Galicia, a physician, wrote to King Manuel of Portugal on 1 May, 1500. Master John had been sent with Cabral as astronomer for the voyage to Brazil. On Monday April 27th he and two of the pilots went on land and took the midday Sun. 'We found 56° and the shadow was [from] north. By this according to the *Regiment of the Astrolabe* we judged that we were 17° distant from the equinoctial, and therefore had the height of the Antarctic Pole in 17°.' The surviving copy of the *Regiment* has a declination of 17° on April 28th, but the calendar is only for a single unnamed year. Master John's procedure was 90° − 56° = 34° (zenith distance of Sun), 34° − 17° (declination) = Lat. 17° S. for observer's position. At sea the pilots appear to have relied entirely on dead reckoning without any instrumental check, for Master John says that their distances all exceeded his, some by over 150 leagues. 'The truth cannot be ascertained until in good time we arrive at the Cape of Good Hope, and there we shall know who goes more correctly, they with the chart, or I with the chart and the astrolabe.' Nevertheless, he had not made much progress with a promised star catalogue owing to the roll of the ship which caused an error of 4° to 5°. 'And I say almost the same about the *Tavoletas de la India*,' which he found very laborious to use. Vasco da Gama had brought home this instrument, the Arab *kamal*, and it is clear that it was considered to have some advantage at least over the astrolabe.

Columbus learned what he knew of navigation during his

years in Portugal, and after his great voyage of 1492 Spanish pilots were naturally eager to master the new techniques. One of the narrators of the second voyage, Dr. Chanca, says quite proudly that some of them knew how to go from Spain and back by the North Star, that is to say by *altura*, while Columbus himself ran down the latitude of (as he thought) Sierra Leone on his third voyage, making anxious observations of the Star with his quadrant every day. A Portuguese manual was translated into Spanish in 1519 by a gentleman who had held office in the West Indies, and following the rule for raising a degree (now shown by means of a star-like diagram of the thirty-two rhumbs) he devotes a paragraph to dead reckoning. Sailors (he says) reckon how far they have gone along the east-west line (the latitude) with the help of the hour-glass, counting what the ship has done each day and each night according to the way it makes for each hour of the glass. And for a good reckoning one must judge by pacing (i.e. with an object thrown overboard) what the ship's way is. But because this is a matter of judgment, the reckoning is uncertain. For safety's sake therefore it is better out of two reckonings to take the highest number of leagues rather than the lowest so that you do not come upon land before you expect it. You should shorten sail and keep a good watch at night. And the same advice holds good for sailing on all the seven quarters —for safety's sake take the greater reckoning rather than the less, shorten sail and look out for land. Only a few years earlier the Italian poet Ariosto had devoted half a dozen lines of verse to a description of dead reckoning, and here he says: 'One on the poop, another at the prow, keeps the hour-glass in front of him, and looks to see every half-hour what distance has been covered, and in which direction. Then each of them with his chart goes amidships and resolves his judgment.' Such a careful procedure was essential for navigation by 'direction and distance' in Mediterranean waters, and on the great ocean voyages there is frequent mention of a number of independent reckonings made by different ships' officers. It was among these that the higher figure was to be preferred to the lower to avoid shipwreck at night, the sailor's greatest dread.

At the very time that the Portuguese were laying the foundations of astronomical navigation, an experienced French sailor,

Pierre Garcie, was writing *Le Grant Routier et Pilotage* for his compatriots, for his Dedication to 'Pierre ymbert mon fillol et cher amy' is dated the last day of May 1483. Garcie's methods are still the traditional ones. The mariner has no chart, but it is assumed that he has lead and line and compass, although these are not discussed. The writer has consulted expert ship-masters from the chief ports of north-west France—Honfleur, Brest, Croisic, Rochelle and others—for his information, and his book held the field for fifty years or so, not only in France but in England, where a translation appeared. The earliest printed edition known is that of 1521, but there may have been earlier ones, and the circulation of hand-copied books continued long after the invention of printing.

Garcie's book opens with the star-clock, giving the midnight position of the Guards for each fortnight of the year and also their position at dawn, or rather 6 a.m. Although, however, a woodcut (in the printed edition) shows the sky-man (Fig. 18), the positions are given in 'ryns des ventz' or wind-rhumbs. For example:

Mid-January, Guards E. at midnight, N. at dawn.
End of January, Guards E. quarter NE. at midnight, N. quarter NE. at dawn.
Mid-February, Guards NE. quarter E. at midnight.

Or to quote from the original:

En la my Septembre, Gardes au suoest quart de su minuyt. Gardes au suest quart de su aube de jour.
A la fin de Septembre, gardes au su quart de suoeste minuyt. Gardes au suest aube de jour.

The new method of using twenty-four out of the thirty-two wind rhumbs for the positions of the Guards introduced an error, for the divisions are no longer in hours as they should be; but precision was not looked for. The book is, however, of great interest because it introduces for the first time actual sketches of prominent landmarks such as were named in the rutter itself. The crudity of these must have been partly due to lack of an expert woodcutter in Rouen, where the French book was printed, and the English printer simply left them out. But it was probably the case that they had long found a place in pilots' notebooks or rather on their tablets which preceded books, and they are used to this day (Fig. 22).

Item, wben you are nortbweſt and by nortb of Uſhant then maye you ſæ thꝛough the poynte whichis to the ſoutb﹐ wartꝭ ofthe maine Jland,and wben you are of of Uſhant nortbweſt and byweſt, then is that poynt ſhutte in on the ſhoꝛe.

Item, wben Uſhant bearesnorth northweſt from you, then dotb it appére like as it is hære aboue demonſtrated.
Item, wben you are offof Uſhant Nortbweſt and by﹐ weſt,oꝛ weſt northweſt then lyes there a great Rocke of the northeaſt pointe, but you cannot well ſæ through be﹐ twixt the Rocke and Uſhant from thence. And alongſt the

The following objects are prominent : Le Stiff twin light-towers (*Lat. 48° 29′ N., Long. 5° 03′ W.*) which stand about a quarter of a mile south-westward of the north-eastern extremity of the island ; the two radio masts, about 1¾ miles south-westward of Le Stiff light-towers ; the light-tower on Pointe de Créac'h, about 1¾ miles west- 15 ward of the radio masts ; the light-tower on Roche Nividic, about half a mile west-south-westward of the western extremity of the island ; and the light-tower on La Jument, a rock situated about 1¼ miles south-westward of the southern extremity of the island.

Île d'Ouessant from west-south-westward.
(*Original dated prior 1917.*)

Lights.—Fog signals.—Radiobeacon.—Distress signals.—A 20 light is exhibited, at an elevation of 226 feet (68ᵐ9), from a circular tower painted in black and white horizontal bands, about 163 feet (49ᵐ7) in height, situated on Pointe de Créac'h, about 8 cables north-

Charts 2643, 2644, 20, 2675a, 2649, 1598, 1104.

Fig. 22. The sea apprentice learnt to draw shore profiles in order to recognize his position. These extracts are from a contemporary Admiralty Pilot (Sailing Directions) and from the sixteenth-century *Safegarde of Saylers.*

The actual rutters in this book are closely similar to those in the medieval English rutter already discussed, but there are added several lists of distances from point to point along the coasts, the unit being the *veue* (the English 'kenning'), or in some lists the league. The *veue* was seven (3-mile) leagues, the kenning about 20 miles, or in Scotland 14 miles. We read, for example, 'Du ras de fontenariba a saint mahe [Matthieu] une veue. De saint mahe au four une veue. De barfleur a cus de caux troys veues.' Yet another novelty was an attempt to apply the Rule of the Marteloio, although the writer soon broke down. He begins: A route of 80 leagues one part (quarter) of the wind outside the route which you want carries a ship 16 leagues 'le hault ou le bas', i.e. presumably 'above or below' it, meaning to one side or the other. A half 'ryn de vent', or two quarters, carries a ship 30 leagues 'le hault ou le bas' in a course of 80 leagues. These two statements correspond approximately to the figures 100 to 20 and 100 to 38 given for the first two quarters in the Italian table. But the Frenchman can go no farther, and contents himself with working out proportionate figures for courses of 20, 40 and 160 leagues, adding that in every case you must know whether the tides are also carrying you 'le hault ou le bas'.

Garcie naturally lays great stress on the tides and the cycle of the Moon but he makes no reference to what is perhaps the most remarkable contribution of north-west Europe to navigation, the Breton tide-charts (Plate X). Five or six examples of these little books have survived, all drawn and painted apparently in the early sixteenth century. But they were the work of professional chart-makers, and, in one copy at least, the chart shows the English flag still aloft at Bordeaux, where it did not fly after 1453. The basic chart (of the ordinary Italian type) runs from Finisterre to Flanders, including the British Isles, and is cut up into four parts, so that Ireland and south Britain (from the latitude of St. Andrews) are mapped each on a separate page. On each map page a large wind-rose is painted with thirty-two points which here represent not direction but the position of the new Moon at high water, assuming it moves a point in 45 minutes. This was in accordance with the accepted way of describing the 'establishment' which has already been noticed: 'Moon north-east–south-west, full sea', for high tide at 3 p.m. The chart-maker must have compiled a list of establishments, for from each port he drew a curving line to the correct point on the wind-rose, from the Thames estuary, for example, to north–south (establishment zero), from the ports of Wales to east–west (6 hours). On separate pages there were eight further wind-roses to represent the different establishment, north, north-north-east, north-east and so on, drawn each in the centre of a large ring which was divided into thirty compartments representing the age of the Moon. Within these compartments, marked concentrically, were first the numbered days of age of the Moon, then sketches of its appearance at the four quarters (or the 1st, 8th, 15th and 22nd days) together with sketches of a sheet of water and a pile of rocks to represent spring tides (3rd and 17th days) and neap tides (9th and 24th days) respectively. The two innermost rings gave the actual times of high and low water in hours and quarters, assuming a daily 45-minute retardation. A daily calendar which included the dates of the Moon's quarters was bound up with the booklet, and by turning in succession to calendar, map and diagram circle a pilot could find all he wished to know about the state of the tide at his port of destination. This neat little tide-manual came very appropriately from Brittany, where tidal difficulties are so severe, and during the reign of Queen Mary an Englishman made a copy to present to the Earl of Arundel, perhaps because of his great interest in the Moon and astrology! But by that time printed calendars could be cheaply bought, and new devices were current for teaching young seamen to 'shift the tides'.

VIII

The Errors of Compass and Plain Chart

IT has been a popular belief that Columbus was the first discoverer of the variation of the magnetic needle, and he was indeed the first man who is known to have mentioned it in his writings. But the fact could not have escaped the attention of the Portuguese pilots who for two generations had observed the Pole Star and reached port by running down the latitude. They said the needle 'north-easted and north-wested' and it was their turn of expression which Columbus used, and used familiarly without feeling any need to explain it as something new. Pilots were not scientists or philosophers, nor did they read learned books; they took things as they found them, so that this behaviour of the needle did not excite discussion, it just had to be allowed for. Some put it down to a poor lodestone or an imperfectly touched or badly hung needle. Others said that it was not because of the needle that the ship appeared to go off course, but because of some hidden leeway. And the truth was obscured besides by the action of the compass-makers who tried to correct the tendency of the instrument to stand a little off the meridian by fastening the wires askew under the fly or card and not exactly under the fleur-de-lis. This explains why Columbus found that his Flemish and Italian compasses did not read alike. While the Flemish, French and some Genoese makers used to make a 'correction' of the needle according to the local variation other Italian makers left it alone, since their customers had never complained. For the Mediterranean method of navigation was by direction relative to a prescribed course as shown on the chart, so that interference was not only unnecessary, but would be positively harmful. For the charts had been drawn, and the sailing directions drawn up, with compasses varying a whole quarter wind ($11\frac{1}{4}°$). There were other craftsmen who had come across the variation and dealt with it also in a commonsense way without comment. These were the sun-dial makers of Nuremberg, who had designed and made a little portable instrument set by a tiny magnetic needle. Finding that the needle did not actually stand in the meridian, they cut a mark across its box to show where it should be when the dial was correctly set. The oldest surviving example of these 'travellers' companions' as they were called carries the date 1451, but they must have been well known twenty years earlier, for Prince Henry the Navigator's elder brother, King Duarte, when describing the sky-clock, mentioned the little *relógios de agulha* or magnetic time-pieces which came from abroad.

The voyages to the Americas and to India between 1492 and 1500 brought a flood of fresh observations of the behaviour of the needle, and particularly of the fact that after a ship had passed the Azores, or alternatively had rounded the south of Africa, there was a change from 'north-easting' to an increasing 'north-westing'. Voyages at this period of exciting discovery were no longer merely the concern of sailors and their employers, they aroused the interest and curiosity of the learned and the cultured world. Accounts of voyages were read, returned pilots and captains were interrogated, and speculation about the variation began. The idea gained ground that there existed a 'true meridian' running through the Azores at which there was no variation, and that the variation increased away from this line, alike to east and west by an equal amount for an equal distance. Here then, it was argued, was a means of checking distance east-west, or, as we should say today, finding the longitude. And without any attempt to check the Portuguese observations in the East (which were not, indeed, made public), the hypothesis was extended to embrace the whole globe. The true meridian would have its counterpart 180° away, and the variation would increase, eastward towards the east, westward towards the west as far as meridians 90° E. and W. from the true meridian, after which it would diminish again to zero. But, speculation apart, it was highly necessary that pilots should be able to ascertain and measure the variation other than by their rough eye-check of the compass at noon, for it threw out their reckoning, and falsified their charting. A foreign physician at the Portuguese court suggested in 1525 that it could be done by setting up a style on a graduated plate with 'a little magnetic needle like those on the sun-dials made in Germany' placed beside the meridian line. And an actual instrument something like this was soon to be in use, although far too narrowly, for pilots were obstinately conservative. And this trait was perhaps confirmed in them (strange to say) by the establishment in Spain of what was to become a famous training school.

It was in 1508 that Queen Joanna of Castile wrote to the then Pilot Major, Amerigo Vespucci, that she understood that Spanish pilots were not so expert and well instructed in the use of the quadrant and astrolabe, and in making necessary calculations, as they ought to be. She therefore laid it down that all pilots should put themselves under instruction, and they were forbidden to ship as pilots, nor were merchant-owners or ship-masters to engage them, until they had received a certificate of approval from Amerigo himself. And he was bidden to give instruction to any who asked for it at his house in Seville. From this beginning the teaching and rigid examination of Spanish pilots gradually took orderly shape in the Casa de Contratación at Seville. The Pilots Major were allowed to appoint deputies and, in 1527, when Sebastian Cabot held office, his deputies were obliged to hold their examinations at the house of Ferdinand Columbus (who in fact knew very little about the subject) to avoid any dispute. Professors and teachers gradually gathered about the school, and each candidate was examined in the presence of a gathering of senior pilots. Charts, instruments and textbooks were also kept under review. By the middle of the century the whole organization was very impressive, and the English Arctic explorer Stephen Borough, who was made welcome in Seville near the end of Queen Mary's reign, later made a powerful appeal to Elizabeth to establish a similar system in England. Perhaps it was as well that she did not, for as the century grew old the Spanish sailing methods became stereotyped and old-fashioned, while in England there was a rush of new ideas.

In Portugal the king had early established a *Casa* or house of Guinea, Mina and India, and here there were repositories of charts, maps and sailing books, and no doubt also of instruments. The quadrant and astrolabe had at some unknown date been supplemented by an instrument that was both simpler and cheaper than either, namely the cross-staff (Plate XIII). Under the name of Jacob's Staff its construction and astronomical use had been described in Latin by the Provençal Jew, Levi ben Gerson, in 1342, and it depended upon the same principle as the old Greek *dioptra*, and the Arab *kamal*. A transom ran to and fro along a five- or six-foot rod, and this rod was graduated to show the degrees subtended at the eye for different positions of the transom. The instrument had its drawbacks: it was for instance more useful for moderately low altitudes than for high ones, and for the stars than for the Sun, when smoked or dark glass must be used to protect the eye. Other observational errors were not noticed in early days and will be mentioned later on in their place, but the cross-staff, the *balestilha* or *baculus*, became a great favourite with sailors, while the quadrant tended to go out of use. Several Portuguese writers mention the cross-staff in the second decade of the sixteenth century, and its prototype had been among the many instruments described and discussed by the German astronomers of the school of Regiomontanus. The great difficulty of securing fine readings was met by the landsman by having large instruments set up on stands or fixed supports. But this was not possible at sea or even when brief landings were made, although Vasco da Gama in 1497 took with him an unusually large astrolabe made of wood which he read ashore. The first device for the subdivision of a scale—the nonnius—was to be invented by a Jewish scholar, Dr. Pedro Nuñez, who became chief cosmographer and keeper of maps and instruments to the king of Portugal in 1529.

It was the king's son, Prince Luiz, who followed in the footsteps of his forbears King John II and Prince Henry the Navigator by interesting himself deeply in navigation, astronomy and the collection of instruments. Both he and his brother, Cardinal Prince Henry, became pupils of Dr. Nuñez, whose mathematical and astronomical writings included two very important tracts on navigation. These were his *Tratado sobre certas duvidas da navegação*, and his *Tratado em defensam da carta de marear com o Regimento da altura*, both published in 1537. The difficulties that he discussed in the first tract are those that every non-mathematical person experiences. How is it that if you turn the ship's head east and make 'straight' for a port due east of you, keeping always at the same distance from the equator, you are not in fact taking the shortest and most direct route? How is it that at the equinox the Sun is seen to rise in the east by everybody in the world whatever their distance from the equator, while at the solstices, for example, people at the same distance from the equator but on opposite sides of it see it rise on the same day in quite different quarters of the horizon? The recognition that the world was a globe was one thing, but to understand the properties of a spherical surface and an 'oblique horizon' was another. The more so as the astronomers declared that the Earth was to be treated as a mere point fixed in the centre of a giant universe. Such intellectual difficulties, however, were of no practical significance. It was the solid 'commonsense' attitude and behaviour of the ordinary pilot and the ordinary chart-maker that led not only to navigational error but to disaster.

The maritime chart was drawn as it had always been, with a circle of wind-roses from which rhumb-lines or quarters of the wind were ruled (Fig. 12). All east–west lines were at the same distance from each other throughout their length, and that was the basis of sailing by east–west *altura*, for throughout its length such a line was also everywhere the same distance from the equinoctial. But the astronomers were saying that the north–south lines did not preserve their equal distances. If two ships head north by the needle they would get closer and closer together, owing to what was called 'the convergence of the meridians'. It sounded sheer nonsense, contrary to all experience. The cosmographers and globe-makers on the other hand had for long been devising various networks for correct plane projections of the sphere, and globes were being made in Lisbon itself for the instruction of the navigator. Furthermore Sacrobosco's textbook, *De Sphaera*, had been translated into Portuguese and was always printed with the navigating manual. But all this was beside the point when a man took out his chart and began pricking his course as he had always been accustomed to do. The chart carried no meridians of longitude or lines of latitude, although round about 1500 some chart-makers had begun to draw a line down the centre and mark it off in *alturas* or degrees north and south of the equator. Such a north–south line appeared, too, on some charts to show how the Pope was considered to have partitioned the world's discoveries, although a commission summoned to Badajoz in 1524 to plot it precisely had broken up in confusion, for there was no way of fixing it over the ocean. One of the delegates had indeed put forward the idea that it could be done by the Rule of Marteloio. Sail out from the Canaries along a certain quarter of the wind, he said, and by the table you can calculate when you are 370 leagues west, and so find where the line is to run. And supposing you found it, how could you mark the sea? they replied.

It was by criticizing an analogous method of measurement that Pedro Nuñez began his 'Defence of the Sea Chart'. Ptolemy had determined the latitudes and longitudes on his map by resolving the direction and distance of a route between two places into easting and northing, using Pythagoras' theorem (Fig. 20). He had also used the same property of the right-angled triangle—the sum of the squares on the sides is equal to the square on the hypotenuse—to find the distance between two places on the globe of which the difference of latitudes and longitudes was known. Johann Stoeffler had written a criticism of this in 1513, on the grounds that it ignored the convergence of the meridians, and that the consequent error in the result became noticeable if the places were more than 18° from the equator. In such a case the difference of longitude in the middle latitude between the two must be substituted for the true difference of longitude. But this use of 'mid-latitude' was not to enter the sailor's calculations for nearly a century. What Dr. Nuñez emphasized was the fact that the direct or great-circle route from one point to another is not a line of constant bearing as it is on a plane surface. And it was now that for the first time the true nature of a rhumb was demonstrated, and in the simplest possible way. Nuñez drew a circle to represent the equinoctial line or equator, the centre point representing the pole. Four diameters represented eight meridians running from equator to poles, and from each of their equatorial points four rhumb lines were drawn so as to cut the meridians at a constant angle. They proved, of course, to be spiral lines eventually terminating at the Pole (Fig. 23). While it is doubtful if any seaman saw or would understand this figure, it delighted the mathematicians and a few years later Gerard Mercator in Flanders drew such sets of rhumbs on his first terrestrial globe.

Nuñez next went on to the errors arising from ignoring the convergence of the meridians, which he put down not to the sailors (for he was at pains to maintain that Portuguese nautical science was the best the world had known) but to the map- and globe-makers. His demonstration is therefore a little perverse,

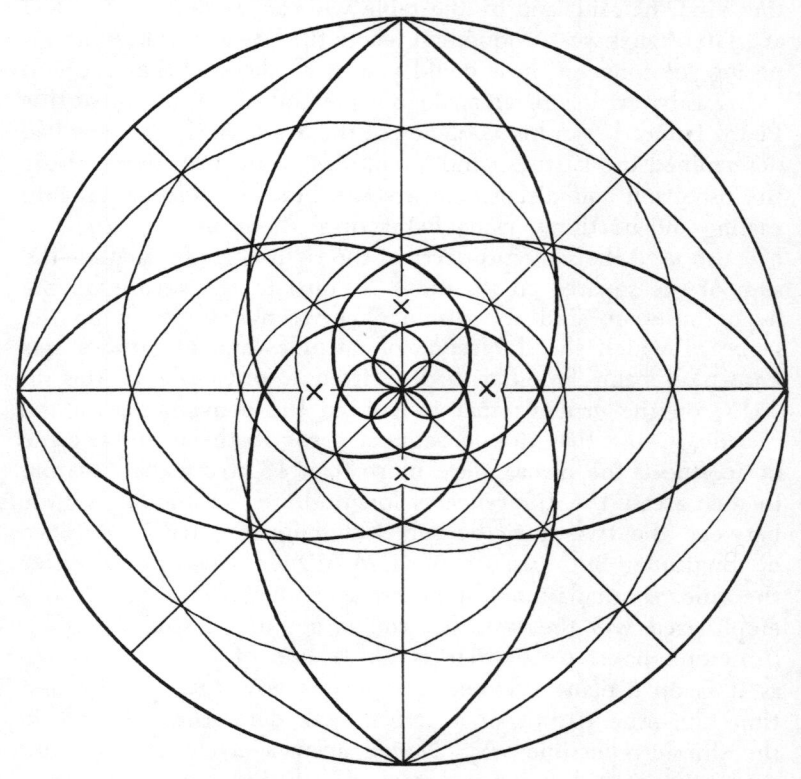

Fig. 23. Pedro Nuñez's diagram of spiral rhumb lines on a polar projection. The lines running ENE. and WNW. from the equator (the bounding circle) have not been carried beyond XX.

since he is defending the maritime chart. He chose his first example from home waters, as voyages to Terceira (Azores) and Madeira are, he says, a matter of daily experience and distances are well known. Terceira is 262 leagues from Cape Caraveyro on the Portuguese coast, and has an *altura* of 40°. Treating all north–south lines as parallel this means that the island is 15 equatorial degrees west (262 ÷ 17½). But 15° in Lat. 40° in fact measure only 201 leagues, and Nuñez draws a diagram to show the converging meridian line. Is Terceira then only this distance from Portugal? No, the voyage has been made too often for such an error of reckoning. But to confirm the figure he will take the position of Madeira. It lies south-west of Lisbon, and 7° farther south (by *altura*). This is equal to 122 (7 × 17½) leagues, and as the direction of the course is 45° from north–south the 'westing' must also be 122 leagues. But Madeira is also south-east of Terceira, and eight degrees farther south by *altura*. Hence the westing of the latter is 140 leagues. The two figures added together make 262 leagues, the accepted westing of Terceira from Portugal. The explanation is, therefore, that the Azorean island is not 15° but 19½° from the meridian running along the Portuguese coast, and this figure is fairly near the truth.

The second example is taken from the India voyage, upon which ships made a great sweep west and then continued on a south-easterly course until they sighted Tristan da Cunha. This island was shown on the chart to be north–south with Cape Three Points in the Guinea Gulf, and 420 leagues from the Cape of Good Hope. But if the meridians were correct, since the island lay in 36° S. it should only be 340 leagues from the Cape, allowing for the smaller size of the degree of longitude. Actually, therefore, it was four or five degrees (in fact it is quite ten) farther west than it was placed on the chart, and was not in the meridian of Cape Three Points. Obviously the sailor and the chart-makers, the map-makers and globe-makers should all take account of the convergence of the meridians and the correct interpretation of 'east–west' in fixing a position. And in a final paragraph of his tractate Nuñez provides the design of a quadrant which will give just this information. The instrument has three scales, one of 90° round the curved edge, one of 17½ leagues along the base (the measure of the 'great' or equatorial degree) and one of one hundred parts on the vertical edge. A silk thread, carrying a sliding bead, was threaded to the corner of the instrument, and a semicircle was drawn upon the base line. To find the length of the degree at, say, Lat. 50°, lay the thread across to this angle, and slide the bead until it lies on the circumference of the semicircle. Now turn the thread until the bead lies on the scale of one hundred parts. It reaches to point x, then $(x/100)$ $17\frac{1}{2}$ leagues is the length of the degree. It will be seen (Fig. 24) that since the angle in a semi-circle is always a right angle the intercept marked by the bead is actually the cosine of the latitude, and so the length required.

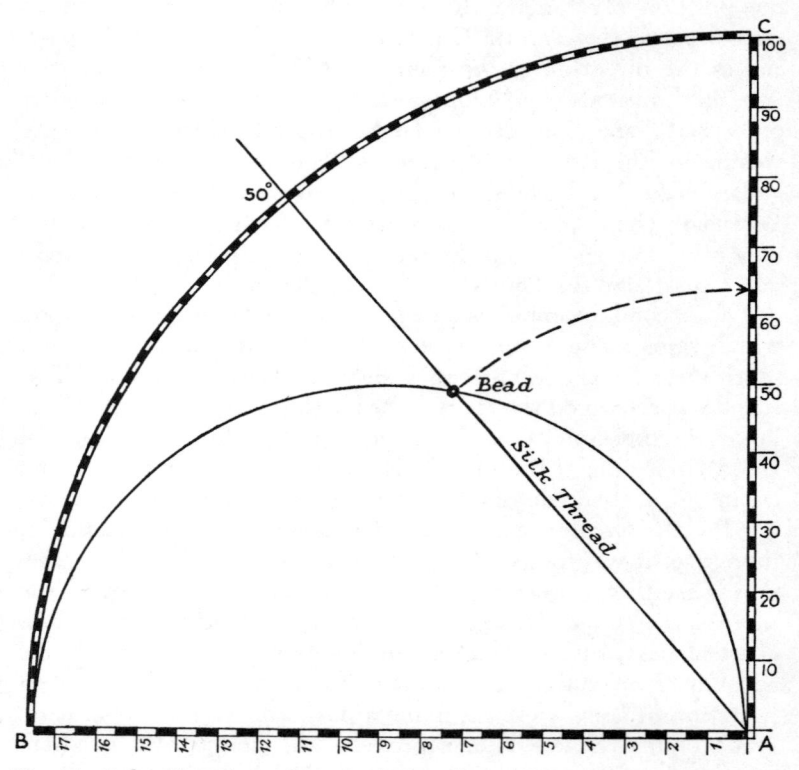

Fig. 24. Pedro Nuñez's quadrant for finding the number of leagues in a degree along each parallel.

Dr. Nuñez had a very poor opinion of the Lisbon globe-makers, whose outlines (he considered) were little more than scribbles although they used plenty of gold paint, and decorated the map with flags, camels, elephants and such like. Of some of the conceited pilots of the India voyage, too, he thought little. They

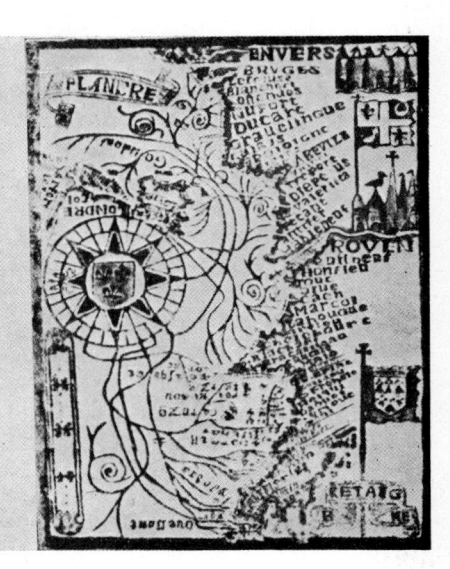

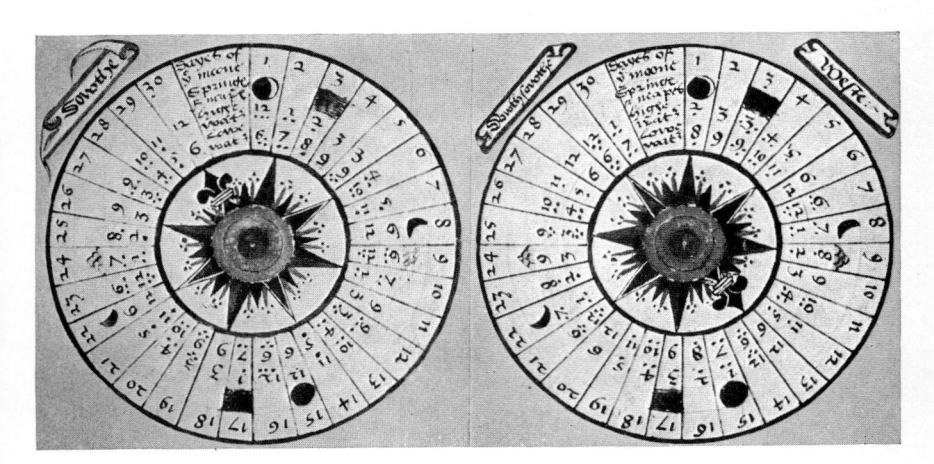

X. In these Breton tide-tables the wind-rose represents 24 hours (S. = noon) from which the hour of high-water at New Moon is read off from a line running to the port. Supposing this is 1.45 p.m. (i.e. SSW.) then from the circular diagram for this bearing (time) the times of high water and low water, and the occurrence of springs and neaps can be read off for each day of the age of the Moon.

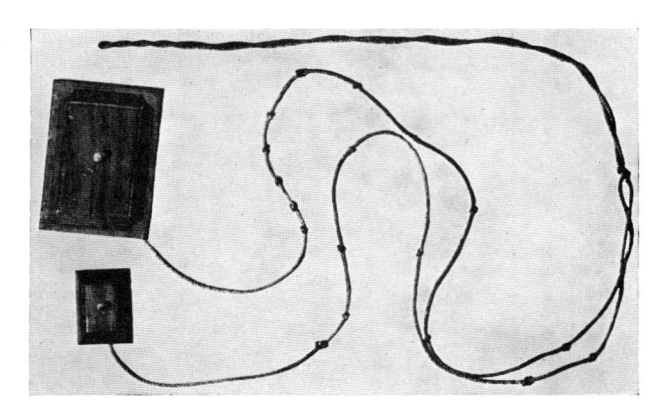

XI. The modern *kamal* has two tablets each with its appropriate range of knots expressing the angle subtended by the tablet according to the distance at which it is held.

claimed, using barbarous language (so he said), to pronounce on the Sun, the Moon, the stars, their circles, motions and declinations, their rising, setting, and bearings; and on the *alturas* and *lenguras* of places all over the globe; and on astrolabes, quadrants, cross-staffs and hour-glasses; on common and leap years, equinoxes and solstices; whereas they knew nothing about them. And then they said that navigation was quite another thing now, they had improved it so much. Some of them even had the impudence to claim knowledge of the sphere (astronomy), and to triumph over those that had not. The learned Doctor had possibly had some very tiresome pupils, while there may well have been something truly painful in seeing the fruits of patient scholarship carelessly handled by uncultured men.

Among his useful notes on the current navigating manual, Nuñez says that declination tables should cover four years, and that it was not necessary to make any correction for the difference of meridians until the user was six hours away from that for which they were calculated, since the readings of the Sun taken with the astrolabe contain greater errors than the small differences that would be found in the corrected figures. This appears to be the first mention of this necessary rectification of the tables during a voyage. The next point to be suggested was unfortunate. Misled by a new but faulty calculation of the precession of the equinoxes made by the German astronomer Werner, the Doctor says that in place of the current correction for the motion of the North Star ($3\frac{1}{2}°$) the figure of $4° 9'$ should be substituted. Actually $3\frac{1}{2}°$ was already too high. However, he advised taking an upper and lower transit which would eliminate any error, although it would not be possible at sea.

He then turns to the important question of the variation of the needle, which he said pilots merely corrected by a glance at the Star, no matter what the time. The instrument which he designed, later called the Shadow Instrument, consisted of a metal plate, crossed by a meridian and an east-west line, and graduated with degrees from $0°$ to $90°$ in the quadrants. A perpendicular style rose from the centre, and a bare magnetic needle was sunk into a circular hollow in the plate, pivoted on the meridian line. The instrument could be hung in cords (an alternative suggestion mentioned later in the book was for gimbals) and the necessity that it should be level is emphasized. The instrument was to be set in the magnetic meridian by the needle, and the position of the shadow marked at the same moment as a forenoon observation of the Sun was taken with the astrolabe by a second observer. This reading was to be carefully kept, and when the Sun descended to the same height in the afternoon the shadow position was again noted. If the mid-point between the shadows coincided with the meridian there was no variation. If it was on one side or the other, then the variation was east or west by half the difference between the two shadow observations. A warning is added that the variation cannot be found correctly in the course of the daily noon observation of the Sun, since experience has shown that owing to the small size of the astrolabe small differences of solar altitude cannot be distinguished, and for some time the luminary appears to move horizontally.

Francisco Faleiro had published a description of a somewhat similar instrument in Spain a few years earlier, but the importance of the one designed by Nuñez rests on the fact that it was used for a careful enquiry into the subject of variation by one of his pupils, John de Castro. This investigation took place in the course of a journey to India in 1538–9. But before describing the really important points then established, it is worth while to run through the rest of Pedro Nuñez's notes, for they all reveal a much higher standard of practice that was now coming into use among the better-trained masters and pilots. He condemns for example the rough-and-ready but faulty way of judging the time (age-old in speaking of the tides) upon the assumption that the azimuths or bearings of a heavenly body changed evenly through the hours. 'When the Sun is in the south-west', he says, 'pilots say that it is three hours after noon, because according to the compass south-west is 45° from south, which makes three hours at 15° to an hour.' And he goes on to demonstrate the inequality arising from the oblique horizon by spherical trigonometry. But then he remembers the non-mathematician, and says that the whole matter can be observed without tables of arcs and sines by means of a simple instrument consisting of two circular plates with a common meridian, one to serve as a sun-dial, the other as magnetic compass (which he calls simply *agulha*, needle). A sketch of the suggested instrument was first published in Nuñez's collected works, which were printed in Latin much later in the century (1566) and had considerable influence on navigation. His mathematical exposition of how to set course by a great circle must likewise have been Greek to the sailor, but here, too, he designed an instrument (published in the later book) by which the successive angles of the course could be taken from a globe. In two subsequent notes the 'shadow instrument' is described with fuller detail, and this is when the suggestion is made that it should be hung in gimbals. It is clear that this method of suspension was already in use for the magnetic compass aboard ship.

John de Castro, who studied the variation, was a famous Chief Pilot of the Portuguese India fleet, and Prince Luiz, who had been very impressed by Pedro Nuñez's methods and teaching, ordered de Castro to use the 'shadow instrument' on his voyage to India in 1538. The prince had a copy of the instrument made by a certain João Gonçalves, said to be famous throughout Europe for his skill as an instrument-maker, and presented it to the Grand Pilot. A brother of Pedro, Dr. Lois Nuñez, travelled with him to assist in the observations, of which a careful record was kept. Entries for each experiment included the time to the nearest five minutes in forenoon and afternoon, the Sun's altitude, and the two positions of the shadow on the scale. Altogether forty-three values for the variation were recorded during the period 1538–41, and de Castro reckoned that a good observer should not make an error of more than half a degree when at sea in smooth water, although on a rolling ship this might increase to 2°. It was soon possible to disprove the hasty assumption of the philosophers that a 'true meridian' of no variation ran through the Azores and Canaries (assumed to be on a north–south line), for in the latter group the needle turned $5\frac{1}{2}°$ and 6° E. at points on the same meridian. Nor were the pilots right who took it for granted that there was no variation at Cape Agulhas. Found to be $19\frac{1}{2}°$ to 20° E. in Lat. $30\frac{1}{2}°$ to the north-north-west of Tristan, it only reached zero at the first promontory of Natal, after which it increased westward up to 11° W. on the coast of India. Some readings were taken by the Sun's amplitude when it rose and set. Dr. Pedro Nuñez had explained in his text that given the Sun's height and declination its amplitude (or angle wide of due east or west at which it rises or sets) could be calculated by spherical trigonometry, and the variation determined therefore by a single observation; but this of course was beyond the ordinary observer of that day, and the first table of amplitudes for use at sea was drawn up by the English mathematician Thomas Hariot.

De Castro had some disquieting experiences, as for example when he checked the Shadow Instrument against three of the pilot's compasses. All the readings differed, and thinking the compasses might have been 'corrected' by their makers, he had them all opened up. But all were 'meridian compasses', that is to say with the wires under the fleur-de-lis, and even re-magnetizing them with the same lodestone did not make them agree. The puzzle remained. On another occasion when making a survey sketch on the island of Chaul he placed his sea-compass on a boulder and the needle immediately turned right round, through 180°. But when he lifted it up it returned to normal, twisting round again every time he put it back. The rock of the boulder itself, he found, was not magnetic, nor could he discover any disturbing object on the island. For he had previously learned on board ship that if there was any iron about (he had been working near a cannon) it set the needles in all directions. This is the first mention of 'deviation' as it is termed, and there was not another for a hundred years, but experience must have taught other sailors that they had to keep iron away from the compass if they wished to maintain course, and a craftsman sees no reason to make such 'tricks of the trade' public. John de Castro, however, with his careful experimental work, was much in advance of his times, as he was, too, in his general attitude to scientific truth, as appears in his declaration when he had to put a new needle into his Shadow Instrument: 'Considering that avowal and honesty are greatly necessary in matters submitted to and under the jurisdiction of the mathematical arts, I declare herein that after arriving in India and reaching Goa the small needle of my instrument was lost.' Unfortunately (as would be thought today) de Castro's report of his observations, like his magnificent rutters, was never put to print. But his writings were incorporated in the manuscript instructions carried by the East India pilots, and it was always said that on this route sailors paid greater attention than usual to observations of the variation of the needle.

Portuguese nautical science was, of course, soon also in the possession of Spain, if only because so many Portuguese pilots and ship-masters offered their services to the rival Crown. The two brothers Faleiro, Ruy and Francisco, went over with Magellan in 1519, although Ruy, who had prepared a Book of Longitudes, lost his reason before the expedition sailed. About twelve years later Francisco wrote an *Arte del Marear* in Castilian, which, after examination by a Professor of Astrology at Salamanca, was licensed for the press in 1534 and given a ten years' monopoly. It is a sound, practical piece of work, beautifully set up and illustrated by the German printer, John Cronberger, who then worked at Seville. And the author was very forthright in his listing of the causes of errors in dead reckoning, putting compass variation in the first place of all, but adding also the imprecision of instruments, and their careless or faulty reading by the pilots, as well, of course, as the unknown currents that caused leeway. He introduces among his definitions the terms latitude and longitude, and gives a neat demonstration of the way in which a pilot who knows the bearing of his port but is in fact east or west of his assumed position will completely miss it while tacking off on the correct rhumb. It is all the more surprising therefore that when he gives the table, or rather two diagrams, of the distance sailed to 'raise a degree' he makes no mention of the corresponding easting or westing made good which was to be found in the earliest manual. He presents two roses of the rhumbs, one supposing that a degree of the meridian measures $17\frac{1}{2}$ leagues, the other taking the figure as $16\frac{2}{3}$ leagues. His declared preference for the faulty Castilian value may perhaps be attributed to his status as a renegade Portuguese.

Within a very few years of the appearance of Pedro Nuñez's treatise, the subject of magnetic variation was taken up in detail again by a French pilot, Jean Rotz, whose father, David Ross, had come from Scotland. It was in 1542 that Rotz presented his treatise, with an elaborate instrument, to King Henry VIII, who had a reputation across the Channel as a great lover of mathematics, as indeed had been true when he was young. The French had become persistent interlopers in Portuguese Guinea and Brazil as well as in Spanish America, and had built up a fine school of pilots and chart-makers at Dieppe. It was to this school that Rotz belonged, although he had lately been studying the works of the cosmographers and astronomers in Paris. As early as 1529 two French brothers had sailed east from Dieppe to Sumatra, accompanied by an astronomer-astrologer Pierre Crignon, who wrote a treatise on navigation five or six years later. This was not printed, and no copy is now known, but an early reader stated that it dealt with the variation, and proposed a method for longitude which was probably very like that which Rotz subsequently expounded to the English king. Rotz had apparently also read the works of Faleiro and Nuñez, but had not, of course, had access to John de Castro's writings, and he believed in the 'true meridian' and a regular increase of variation east and west of it. But he reckoned the relationship in leagues of distance and not in degrees as the more general theory demanded. His calculations afford a useful illustration of the general tendency of the age to argue from insufficient observational data, for he was satisfied with a list of four variations from which he deduced that the needle altered one degree in every $22\frac{1}{2}$ leagues, whatever the latitude. Hence if on his outward voyage the sailor observed the variation, as well as the latitude, he could find his position with certainty on his return, for the difference of variation would give him his easting or westing.

The instrument that he proposed for the observations, and of which he brought a prototype to the king, looks very clumsy in the sketches which he drew of it, but these may be misleading. Briefly it consisted of a set of what were termed 'astronomer's rings' attached so as to turn in the vertical plane above a large open-needle magnetic compass. The rings included an equinoctial and a meridional circle, the latter furnished with pinhole sights so that the instrument could be set from observation like a globe, with the axis (represented by a thread) corresponding in direction to the axis of the Earth. The general principle was that used also in Dr. Nuñez's instrument, namely the comparison of the Sun's true azimuth as shown by a shadow with the azimuth determined by the magnetic needle. But it would have taken an extraordinarily skilful instrument-maker to make Rotz's Differential Sun-dial (*Cadrans Differential*) as he called it, read correctly, and in fact there is no suggestion in the accompanying tract that he ever made any actual observations with it. He was a professional pilot and chart-maker seeking patronage, and it was the convention that a man should present as his testimonials some example of his skill. In this case the application was successful, for the Frenchman was appointed Hydrographer to the King at a salary of £40 a year. But his treatise on navigation was not 'put to print' as he had hoped.

Much of the text was concerned with the elements of astronomy, with the ordinary rules for finding latitude by Sun and star, and even with the elements of land surveying. And the current marine chart with its rectangular north–south and east–west lines was attacked not on the grounds of the convergence of the meridians on the globe, but because except on the 'true meridian' the north–south lines varied from north. Nevertheless much light is thrown incidentally upon current thought and practice, as when he attacks the still persistent argument that it was leeway, and not magnetic variation, that had to be allowed for in crossing the Atlantic. A pilot who set his ship's head due west found himself continually getting nearer to the equinoctial, and to sail down the latitude of, say, Deseada, it was necessary to set course west-a-quarter-north. Currents, said the pilots, set the ship south. But, says Rotz, on the return journey the ship sailing due east by the compass finds herself to the north of east. Is the current then reversed? Latitudes, too, set merely by the compass course did not agree with those set by Sun and star, the most notorious case being in the eastern Mediterranean, where the traditional chart, if a scale of *altura* was added, showed Venice in 50° N. instead of 45°, Rhodes in 42° instead of 36° and so on. This was countered by putting a different scale of degrees on the eastern and western margins of the map, a device Rotz himself employed on several sheets of the atlas of charts which he drew for Henry VIII. And he said that on some Spanish charts three such scales, a central and two marginal, were used. But at the same time there were pilots who carried compasses corrected for the variation for each part of the voyage—Rotz declares he knew a fleet that set out with six differently corrected instruments. And since the Dutch chart-makers were now beginning to draw lines of latitude across charts (as they had long been drawn on maps) the confusion can be imagined. It was essential to make observation of the variation a matter of daily routine, and this was gradually to become common practice later in the century. And it is worth noticing that long before Jean Rotz's day, in 1504 as a matter of fact, the variation on the Grand Banks, reckoned as two points ($22\frac{1}{2}°$) at Cape Race, had led chart-makers to add an 'oblique meridian', i.e. one showing a true north–south line in relation to the compass rhumbs, which was divided into degrees or *alturas*. The sailors of Galicia, Brittany, the Azores (and even some from England) had flocked to the teeming cod-fisheries reported by John Cabot and the Portuguese explorers, and when the French entered Canada they used this same device on their maps.

In addition to his suggestion of how to find the 'east–west' position by variation, Jean Rotz gave some elaborate (but fictitious) examples of how it could be done by lunar distances, for he was a diligent reader of the German astronomers, including Stoeffler and Peter Apian. The latter suggested, as Werner had earlier done, that the cross-staff could be employed for the necessary observations, and Rotz's example of using it (translated from the original French) runs as follows:

'Being at sea and wishing to find the difference of my meridian and that of Dieppe, I rectify my Ephemerides or Alphonsine Tables for Dieppe, and then find the true place of the Moon for the coming night, say 10 o'clock, which is 16° in Taurus. To find its declination, it is necessary to subtract its latitude, which is S. 5° [from the ecliptic] from the declination of the point 16° in Taurus, which is "about 16°", giving 11° as the declination of the Moon. Next I take a star, which is *Aldebaran*, or Oculus Tauri, and find (after all rectification) that it is in Gemini 2° 15', which has a declination of N. 21° 5'. Then I find the latitude of *Aldebaran*, which is S. 5° 10', and hence by subtraction its declination is 15° 55'. And note a point here, you must rectify your time from the Equation of Days, taken from the Alphonsine Tables.

Now to find the true distance between the Moon and the star, I take the longitudinal difference, which is 16° 15', and also the difference of their declinations, which is 4° 55', and for greater ease in getting square roots I first reduce them to minutes. Then multiplying each one by itself I get for the first difference 956,484 minutes and for the second 87,025, which added together make 1,043,509 minutes. Taking the square root gives me 1021 minutes or 17° 1', which is the true distance between Moon and star. [This was the standard method of finding distances between points either on the celestial sphere or on the globe.] Next I divide them [17° 1'] by the minutes which the Moon travels in an hour on the said day, which are 30', and so 34 hours 2 minutes will be the time in which the Moon will be in conjunction with the said star. Consequently tomorrow at 8 hours and 2 minutes after midnight the Moon will be in conjunction with the star at the meridian of Dieppe. And the said hour for which I have made my observation having come [i.e. 10 p.m.] I take the distance of the Moon and star with the cross-staff and find it 12°.

And as it is less than the distance found for the meridian of Dieppe, and the Moon is west of the star, I say that I am west of Dieppe. Then I divide the difference, that is to say 12°, by the movement which the Moon will have made on that day in an hour, which is 30 minutes, and this gives 24 hours as the time in which the Moon will be in conjunction with the star. And thus tomorrow at 10 o'clock at night the Moon and star will be in conjunction at my meridian, and at Dieppe they will be in conjunction 8 hours and 2 minutes after midnight. So that I say that the conjunction will be 10 hours and 2 minutes earlier at my meridian than at Dieppe. Reducing this to degrees by multiplying by 15 gives 150° 30′ that I am west of Dieppe.'

Rotz had been in Guinea and in Brazil, but certainly never so far as $150\frac{1}{2}°$ west of Dieppe. But the imperfections of the Ephemerides, of his method of finding the 'true' distance between Moon and star, and of his estimate of the Moon's motion, combined with the crudity alike of the observation and of contemporary timekeeping, sufficiently explain why the method, so seemingly simple, was to remain merely theoretical for a century and more. Columbus had thought that he found his longitude by noting an eclipse of the Moon of which the time had been foretold in Regiomontanus' Ephemerides, and Amerigo Vespucci had attempted a lunar distance. But the results they obtained were quite worthless—worthless that is to say as determinations of longitude, but worthwhile as pointing the need for instruments of precision to bridge the gap between theory and practice.

When King Henry VIII died, Jean Rotz bribed his way back to France by a promise of English harbour plans and maps. But by that date the English mathematicians and sea-captains were preparing to make their own contribution to navigational practice, a story to be told on a later page. Meanwhile, it is worth glancing back at Spain. In 1545 her Cosmographer Royal, Pedro de Medina, produced a handsome manual, or *Arte de Navegar*, which because of his nation's prestige was translated into both French and English, although the writer held, for example, that compass variation was merely due to the circular motion of the Pole Star. His successor, Martin Cortes, wrote an even fuller and more efficient manual in 1551 under the same title, which went deservedly through many editions and was translated into many languages. It contained nothing that was really new though it reveals that the Spanish chart-makers dealt with the trouble of the parallel meridians by spacing them correctly, not for the equatorial degree, but for the roughly middle latitude of Cape St. Vincent, 37° N. This, of course, lessened although it did not remove the errors arising from pricking a course on such a chart. It had, however, the authority of Ptolemy, although he suggested it only for relatively small areas. But Cortes supplied also a very full table of the true length of a degree at each different latitude, expressing it as a proportion of the equatorial or 'great' degree using angular measure. Taking 60 equatorial degrees as a unit, their equivalent in Lat. 1° is given as 59° 59', in 2°, 59° 58', and so on. Thus in 15° N. or S. the equivalent length is 57° 57', in 45° N. or S. it is 42° 26', in 60° N. or S. 30°, in 90° N. or S., 0°. And the advantages of this table were twofold. The teaching of arithmetic or 'algorism' always included the Golden Rule, i.e. simple proportion, and by giving these ratios the pilot was left to use whatever linear measure of the 'great degree' seemed to him the correct one. In particular he could choose between $17\frac{1}{2}$ leagues (which Cortes appeared to favour) and $16\frac{2}{3}$ leagues. Suppose, for example, he wished to turn his easting into degrees of longitude in latitude 45° N. The proportion is: as 60° is to 42° 26' so is $17\frac{1}{2}$ leagues (or $16\frac{2}{3}$ leagues, or 60 miles) to the length of a degree at this latitude. The rule for the resolution of the course into northing and easting is given in the same terms. For example, to raise one degree of *altura* if sailing on the first rhumb, you must cover a distance equivalent to 1° 1', making an easting equivalent to 12'. Sailing on the sixth rhumb you must cover a distance of 2° 37', when the easting will be 2° 25' (in 'great' degrees). The 'great' degree was of course a degree of a great circle, but here it was merely a device for expressing a length, and the difference between a rhumb and a great circle course was not raised.

To a modern reader familiar with the language of elementary mathematics, the directions given to the pilot in these early manuals appear intolerably long-winded, and their circumlocutions difficult to follow. Here, for instance, is the English translation of Cortes's directions to the pilot for pricking the chart after an observation for *altura* has been taken:

'If he find himself in more or less degrees let him take two pair of compasses [i.e. dividers] and put the foot of one in the point or place where his ship was when he departed, and the other on the line or wind by which he saileth; and likewise let him set the one point of the other compass in the graduation of the card in that number of degrees that he findeth the latitude of the pole, and the other point of the same compass in the next line of east and west: and so with both the compasses, one in one hand and one in the other hand, let him go joining them together taking good heed that the point of the [one] compass do not swerve from the wind whereby he sailed: neither the point of the other compass from the line of east and west where he set it. And following these two compasses by these two lines until the points of the two compasses join . . . then where these two points join is the point where the ship is.'

In point of fact the position of the ship is where the line of the rhumb followed intersects the line of the latitude observed. But on the plain chart, which in fact distorted direction and distance alike, it would have been a difficult matter to state that position in terms of the globe—or in terms of the nearest lee-shore.

IX

The English Awakening

ONCE ships had become so large that they usually lay at anchor instead of being drawn up on the beach the ship's boat became a necessity. And from the beginning of the Great Age of Discovery it was a matter of routine, when some strange bay or river-mouth was reached, for the master or mate to take the skiff and row in, and so by casting the lead search out the channel and lead the ship to her anchorage. It was a matter of routine likewise for one or more of the ship's officers to go ashore, climb the nearest eminence, and with the help of the compass make a sketch-plan of the coast. This supplemented the shore-profile, landmarks, bearings and depths which were all taken from seaward. When a rutter was transcribed for some noble or royal patron these rough materials were worked up into a finished form by a professional painter, as can be seen in the case of the fine illustrations which decorate the rutters of John de Castro. He, it will be remembered, was himself making a sketch-map from a hill when his compass needle behaved so oddly. An earlier illustrated Portuguese rutter had been included in a work on cosmography and navigation written about 1505–8 for King Manuel. The author, Duarte Pacheco, was a gentleman who had sailed frequently to Guinea and to India in the royal service, and, advising the pilot to study the numerous picture-plans carefully, he says that these were 'painted from sight'. Unfortunately the copyist did not include them in the only surviving manuscript of the work. Such harbour-plans were of course of great importance also for military operations in home waters, and a number of them, part picture, part map, were made for King Henry VIII of the English Channel ports. These were presumably for use in his French wars, and had been drawn by his military engineers, but they mark the beginning of real maritime survey in this country, involving as they do some knowledge of geometry and perspective.

The oldest surviving English maritime chart was drawn by a young gentleman, Richard Caundish of Suffolk, whose qualification was his skill in geometry, then a most unusual accomplishment in England. The chart was of the Thames Estuary with its complex sandbanks, and the gifted young man was employed in 1540 to supervise the harbour works at Dover, and make a scale plan of that port. Associated with him were two sailors who seem to have been almost the only ones in the country who had attained the technical skill in sea practice which the new century demanded. One of them was a Thames pilot, who had not only brought the king's ships safely up the river but had found out a new and better channel through the Black Deeps. For this he was rewarded, and it may be presumed that he had searched out the channel with lead and line and subsequently either charted it, or supplied a description which other pilots could follow. The second man appointed to work with Richard Caundish was a ship-master in the service of Viscount Lisle, who was later to become Lord Admiral, and still later that ambitious Duke of Northumberland who could not brook the maritime supremacy of Spain. John à Borough is first heard of in 1531 sailing to La Rochelle, and again writing to his patron from Venice, but it was only because he became involved in an action at law that we learn that he had mastered the Spanish technique, or perhaps it would be truer to say the Portuguese technique, of seamanship. He was suing for the value of his two sea-chests, lost in 1533 when he was master of the *Michael* of Barnstaple. Their contents included three ships' compasses, worth 23s. 4d., of which two were said to be 'otherwise called *carakake*', presumably a Spanish sailors' name. There were also a balestow (cross-staff), and a quadrant, instruments hardly known in England, which together with a lodestone and a running glass (sand-glass) were valued at 26s. 8d. The chart was one which 'contained and served for all Levant', that is to say for the whole Mediterranean Sea, and à Borough had two rutters, one in Castilian, the other in English 'which I John à Borough was a year and a half of making of it'. Finally there was a Reportory in Portuguese which can only have been the *Reportorio dos Tempos* of Valentim Fernandez, first printed in 1518, which went through several editions, and contained, besides calendrical material, the rules and tables of the early nautical manuals.

The appointment of this particular trio—a scholar, a pilot, and an up-to-date ship-master—to supervise the king's works at the principal Channel port is of interest in itself as showing that the need was felt for the combination of theory and practice. But three instructed men do not equip a nation—nor even would half a score. One Richard Hall is to be heard of, getting 5s. over his wages for 'pricking a card' of a voyage made in 1539; a John Rut took his ship to the West Indies and Newfoundland as early as 1526; and a Roger Barlow tried to persuade the king to adventure to Cathay, by presenting him with a table of solar declinations and a *Brief Summe of Geographia* from the Spanish. There was, too, a Scots pilot, Alexander Lyndsay, who conducted James V by sea all round his kingdom, and prepared a rutter after the old customary style with the emphasis on tides, depths and the sea-bottom. But the English king had to look overseas for competent pilots to man the navy, as did those making new adventures across the oceans, and the French Ambassador remarked in 1540 that English ships were full of Ragusans, Venetians, Genoese, Normans and Bretons. When the *Barbara* went to Brazil in that year she had a French pilot, and in the lawsuit which followed he is stated to have had 'an Astrolaby, a Balestely and an instrument belonging to the office of a pilot for the night'. The English witness did not even know the Nocturlabe or Nocturnal (Fig. 19) by name, and the foreign *balestilha* had yet to become the English cross-staff. When Lord Lisle was appointed Lord High Admiral he engaged Jean Ribault, a noted French Huguenot pilot, at a generous salary, and a few years later brought Nicholas de Nicholai, a skilled chart-maker, over from Paris. Soon after Henry VIII died there were said to be three-score French pilots in England, many of them standing high in their profession, and the new French king, Henry II, ordered them all back. Nicholas the Painter (as he was called) salved his conscience, as Jean Rotz did, by taking back numerous English harbour plans and maps, including a copy of Alexander Lyndsay's Scots rutter, and the map that accompanied it. Jean Ribault was kept in the Tower, and as Sebastian Cabot had just returned to England the two were set to work on the material, such as charts, sailing directions and other equipment, which would be necessary for carrying out the Duke of Northumberland's remarkable plans for voyages outrivalling Spain. Not only the Frenchman and the Chief Pilot of Spain, but a Portuguese gentleman-pilot at odds with his king, 'the noble Pinteado', was called into these conferences—evidence enough of the lack of English maritime skill.

But English brains and English talents were not wanting, although these had to be redirected in half a score of new ways before the English mariner could become the equal of his Portuguese, Spanish or French contemporary. And it was only another fifty years before he began to outstrip them all in methods and inventions. In the first place it was necessary for a certain number of mathematicians and astronomers to consider sailors' needs, and to improve the theory of navigation, besides compiling more accurate tables. It was necessary, too, for vernacular textbooks to be written on elementary mathematics for the seaman's use, as well as textbooks on navigation. And this carried with it the need for instructors and teachers of the theory of navigation, which could not be picked up aboard from masters and mates of the old-fashioned type. Moreover the new methods involved the use of instruments, so that a new class of mathematical instrument-makers had to be brought into being to fill the demand; while all these changes needed fostering by the interest and patronage of men in the governing and wealthy classes, who in their turn would be influenced by government policy.

The new voyages that were begun after 1550 to Barbary, Guinea and the north-east had in fact the enthusiastic support of the court, and the Duke of Northumberland's son-in-law, Sir Henry Sidney, made himself responsible for the training of a young Bristol seaman, Richard Chancellor, as chief pilot for the Arctic. The theoretical and astronomical training of Chancellor was put into the hands of John Dee whom Sidney and the young Dudleys (Northumberland's sons) had heard lecturing in Paris to a huge and enthusiastic audience on the geometry of Euclid. Dee, in his turn, had been deeply influenced by the months he spent at Louvain after leaving Cambridge University. There he found the Professor of Mathematics, Gemma Frisius, and his pupil Gerard Mercator, as deeply interested and engaged in the designing and making of mathematical instruments as in the theory of their subject. And in this they had the patronage of the Emperor

Charles V, who was also king of Spain, although he kept his court at Brussels. Dee took some of their instruments and globes back to Trinity College, of which he was a Foundation Fellow, and side by side with his own great mathematical library at Mortlake he built up a collection of mathematical instruments and taught their use. The first London professional instrument-maker, Thomas Gemini, came also from the Low Countries, and had set up business in Blackfriars as an engraver, but from 1553 onwards until his death in 1562 he also made astrolabes and surveying instruments. With his successor (possibly his apprentice) Humfrey Cole (1530?–91) the English mathematical-instrument-making industry was fairly launched. Meanwhile Chancellor made some of his own instruments, and for the giant quadrant (5 ft. semi-diameter) with which he and John Dee 'took the Sun' for new declination tables, he had independently invented the diagonal scale. Dee also had a great cross-staff or *radius astronomicus*, 10 ft. long and mounted on a stand, which had the scale subdivided in the same manner.

Three years later, after he had opened the White Sea route to Muscovy, Chancellor was drowned at sea. His work and methods were followed up by two brothers, Stephen Borough who had sailed with him as master and William who went as apprentice before the mast. These two in their turn became the pupils of John Dee, and Stephen's Journal shows that in 1556 they were competent in taking the Sun and finding the variation of the needle. 'I went on shore and observed the variation of the compass, which was $3\frac{1}{2}°$ from the north to the west. The latitude this day was 69° 10'. It higheth on the barre of Pechora 4 feet water, and it floweth there at a S.W. Moon a full sea.' In default of professional chart-makers, too, William perfected himself in this art, and made a 'plat' for the northern navigations, besides other charts of which one or two have survived. It was for the correction and extension of this 'plat' that the Bassendine voyage of 1568 was intended, and although in fact it never sailed Borough's instructions for the coastal survey were preserved in Hakluyt's *Voyages*. But it was Stephen Borough who, after the Arctic explorations of 1556, had been invited to witness the now famous examination of pilots at Seville. He tried in vain to get the profession systematized in the same way in England, but he did at

XII. The traverse-board (found in the Isle of Barra, 1844) goes back to the sixteenth century, and was used by the helmsman to peg the half-hours run upon each rhumb of the wind.

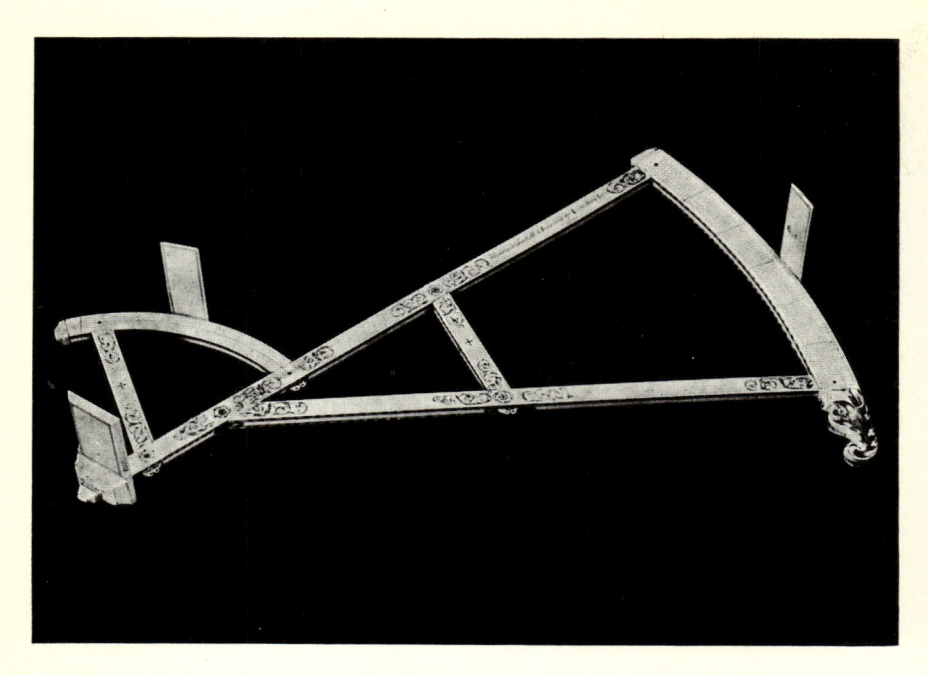

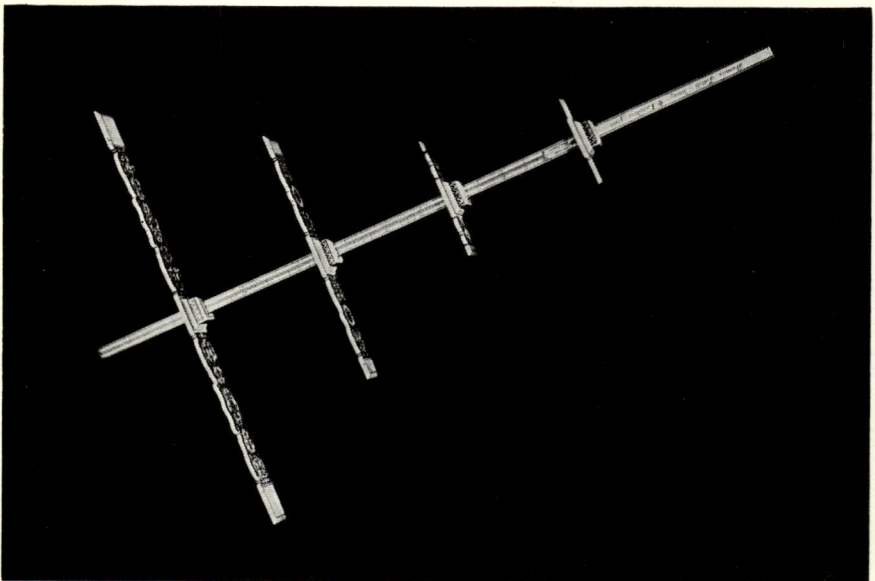

XIII. These presentation instruments (back-staff and fore- or cross-staff) were made in ivory by Thomas Tuttell (*fl*. 1695–1702). The four cross-pieces are intended to be used separately, the staff carrying four scales.

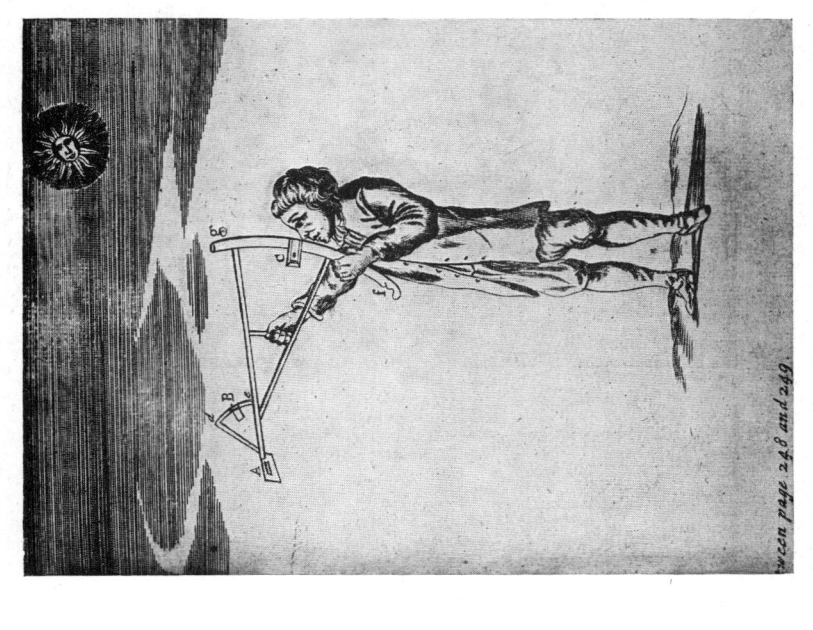

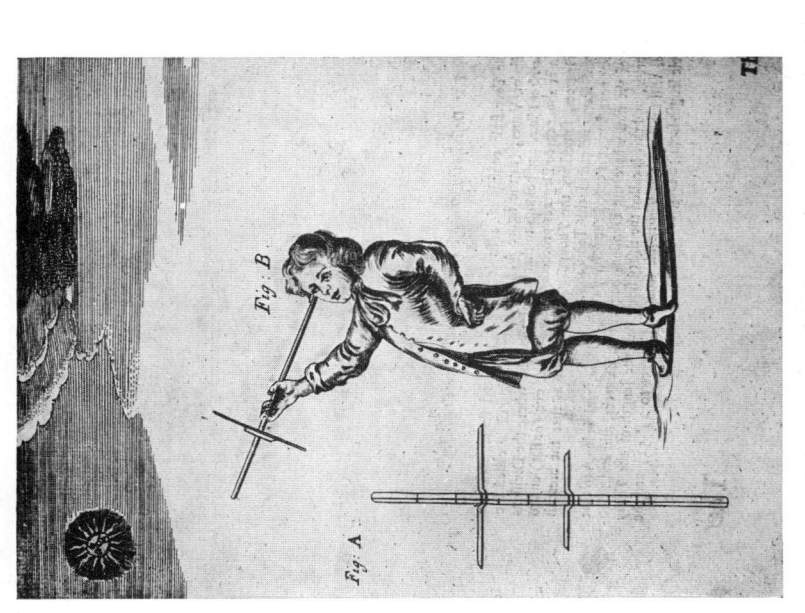

XIV. The two Christ's Hospital boys show how the fore- and the back-staff should be held (1681). (From Jonas Moore: *A new System of the Mathematicks*.)

XV. A sixteenth-century chart based on rhumb-lines. Its Portuguese origin is indicated by the heraldic castle. As the magnetic variation is disregarded the Eastern Mediterranean appears too far north.

least succeed in persuading a group of London merchants to pay for the translation and publication of a Spanish nautical manual, Martin Cortes's *Arte de Navegar*. This was in 1561 and the translator was Richard Eden, the Cambridge scholar who had chronicled the first English voyages to Guinea. Dr. Recorde, a mathematician who had promised the Muscovy Company to write an original English textbook on navigation, had died before he could carry out his purpose, but his series of vernacular textbooks on elementary arithmetic and astronomy, begun in 1543, proved of great service to the new type of sailor. John Dee, unfortunately, preferred to keep his specialized knowledge out of print, and it is impossible to tell quite what his new instrument was which he called the Paradoxall Compass, and taught the use of to the Boroughs. It was not, of course, a mariners' compass, but appears to have included a polar zenithal chart for setting course and avoiding the errors of the plain chart. Dee was a great admirer of Pedro Nuñez, and he certainly taught the true nature of rhumbs, and the method of finding an approximate great circle course as the Portuguese scholar had done.

Another intellectual contributor to English seamanship was Leonard Digges, a close friend of John Dee's, whose career was unfortunately ruined when he joined in Wyatt's rebellion. Digges had a strong belief that the artisan should be introduced to elementary mathematics and instrumental measurements in place of rule of thumb, and he wrote and taught accordingly. His most popular work was his so-called *Prognostication*, a calendar with much supplementary information which from 1553 onwards went through a great number of editions. It included a tide-table giving the establishment of thirty-seven British and Channel ports expressed in rhumbs of the Moon, besides instructions how to calculate the time of high water given the age of the Moon. There was also a table of the Sun's altitude for each hour throughout the year at Lat. $51\frac{1}{2}°$. This allowed a sailor to find his difference of latitude from London in the way the very oldest nautical directions discussed on page 161 had suggested. Other tables were those required for graduating cylindrical dials, astronomers' rings and other instruments. A reader of Gemma Frisius's works, Digges included in 1556 a description of his adaptation of the geometrical square for sea-reckoning, terming it 'an Instrument for Navigation most commodious', although its use was only to be taught to his private pupils. The square in its common form was usually engraved on astrolabes and quadrants, and served as a graphical means for determining (in a rough way) the tangents and cotangents of angles—termed *umbra recta* and *umbra versa*, presumably because they were first used in taking the height of the Sun by its shadow. The principle of the square was simple (see Fig. 25). Each side was marked off in twelve divisions from top to bottom and from left to right. A line of sight to an object elevated 45° above the horizon would fall on the corner, where the ratio of the two scales (the tangent of the angle) was 12 to 12 or unity. If it fell at a less angle, say on scale point 5, then the

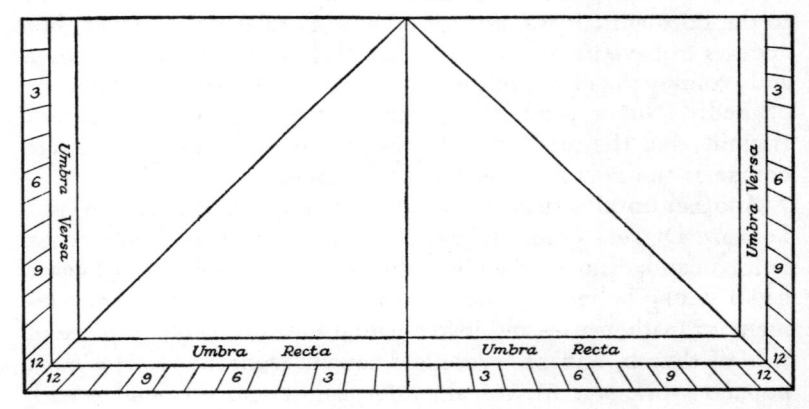

Fig. 25. The Geometrical Square.

angle was defined by the ratio 5 to 12. A line of sight at an angle greater than 45° would fall on the bottom horizontal scale, and if, say, at division 5, then the tangent of the angle was 12/5. The ratios would, of course, also be those of a gnomon or style to its shadow, defining different angles of the Sun; and ever since the days of Leonard of Pisa a stereotyped series of examples had demonstrated how such ratios could be used to find the height of a distant object if its distance were known, or its distance when viewed from two points a measured distance apart. These illustrations of the properties of similar triangles which formed part of every mathematician's textbook on the astrolabe were, of course, merely theoretical examples or exercises, for the ordinary student had no knowledge of, nor interest in, the practical problems of land-survey or navigation. It was exceptional men like Gemma Frisius, John Dee and Leonard Digges who turned these mathematical propositions to the use of the professional engineer and the seaman, the instrument-maker and the artisan. The Nautical Square was made up of four quadrants or quarter squares, with eight rhumbs in each running out from the centre. The graduations along the edge gave the proportions of northing and east—westing for each rhumb, although Gemma Frisius, like Leonard Digges, kept the detailed explanation of the use of the instrument to himself.

By the early years of Elizabeth's reign, it could be said that there was a lively intellectual interest in the need for the advancement of the technique of navigation. Dr. William Cuningham, for example, had dedicated to Robert Dudley a book, the *Cosmographical Glasse*, which devoted a long section to what he called hydrography, while he dealt, too, with simple surveying. But books written by scholars, even if written in English, were heavy going for the man who had barely mastered the three R's, and so far no one had written for the sailor in his own language. Stephen Borough's summing up of the needs of the situation had lain before the queen's advisers in the New Year (January 3rd) of 1563, and although there is no doubt that his two papers constituted a special plea for his own appointment as Chief Pilot of England, they contain many points of more general interest. In Spain, he said, the order of degrees on board ship was as follows: pilot, master, mariner, gromett, page and boy, whereas in England there was only man and boy, a boy becoming a man for his age and not for his knowledge—a rather unfair representation of the apprenticeship system. Moreover, most English masters were satisfied with 'the old ancient rules' as they termed them, while even those ready to learn found that those who had the new knowledge would not teach them, for fear of hindering their own living. Many, indeed, now took instruments to sea although they were ignorant of their use, but were deterred by shame from acknowledging that this was the case; or perhaps they could just take a latitude, and then considered that they had climbed to the top and knew all. It often happened that when two or three ships were in company the masters gave totally different opinions of their distance from land. And this was not due to defects in the art of navigation 'which is certain', but to ignorance of the art. Heavy losses of ships, such as had recently occurred off Spain and Brittany, could be prevented by the appointment and authorization of a man learned and skilful in the art of navigation to teach and instruct the ignorant, and to see that only competent men exercised the pilot's office. And Borough goes on to praise this wonderful art by which 'with a compass and certain lines drawn

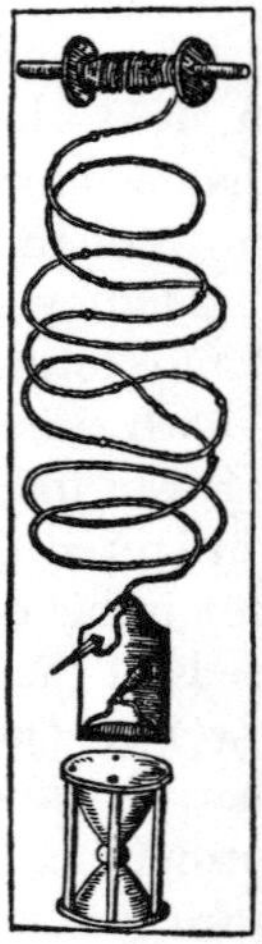

Heures.	Nœuds.	Brasses.	Routes. Rumbs.
2	3	2	Cap au Nort ¼ du Nordeſt.
4	2	4	Cap au Nort-nordeſt.
6	4	2	Cap au Nor-deſt.
8	5	3	Cap au Nor-deſt.
10	2	3 ½	Cap au Nort ¼ du Nordeſt.
12	3	5	Cap au Nort-nordeſt.
2	2	3.	Cap au Nordeſt ¼ de l'Eſt.
4	2	4	Cap au Nor-deſt.
6	6	1	Cap au Nort.
8	6	3	Cap au Nordeſt ¼ du Nordeſt.
10	6	2	Cap au Nort ¼ du Nordeſt.
12	3.	4	Cap au Nort-nordeſt.

Fig. 26. A seventeenth-century log and line and sand-glass.

upon parchment a man can direct his course to places that he cannot see'. With an astrolabe no broader than one's hand the vast heaven could be measured, and the height of the huge Sun through a tiny hole. The mariner's compass, too, which 'gives light and certainty to navigation' was nothing more than a piece of paper—the fly—of a palm's breadth, with a piece of wire fastened beneath it, which moved of itself by virtue of the lodestone.

No appointment of Chief Pilot and Instructor was made, but a few years later a citizen of Gravesend (mentioned as an innkeeper and port-reeve) stepped into the breach. William Bourne had read the books of the learned world, and had also known the sailors of Gravesend intimately. He decided that to bring the two together something in more homely style than Eden's *Arte of Navigation* was needed, and in 1567 published *An Almanack and Prognostication for iii Yeres with Serten Rules of Navigation* which picked out from Eden's translation of Cortes the essential points of the navigator's art. Bourne gave personal instruction besides, and after his *Serten Rules* had gone through its second three years' run he launched a full-scale manual called *A Regiment for the Sea* (1574) which became the English seaman's vade-mecum for a generation and more. It was translated, too, into Dutch—a tribute to its merits.

The *Regiment* included, besides the routine for observations of Sun and star, a description of the first original English contribution to the mariner's art, which apparently had been in use long before this date. It was the log and line (Fig. 26). The crux of navigation was to be sure of the ship's way, and more than one instrument-maker thought he had found a means to measure it mechanically with a geared instrument. The French king's engineer, Jacques Besson, described one, and Humfrey Cole made something similar. But they did not work, they were much too elaborate. And what the English sailor did was to tie a log of wood on to a cord, heave it overboard, and measure the line paid out while someone turned the minute sand-glass.

> 'To know the ship's way [writes Bourne] some do use this, which (as I take it) is very good: they have a piece of wood, and a line to vere out over boorde, with a small [i.e. thin] line of a great length, which they make fast at one ende, and at the other ende, and middle, they have a piece of lyne which they make fast with a small thred to stande lyke unto a crowfoote: for this purpose that it should drive asterne as fast as the shippe doth go away from it, always having the line so ready, that it goeth out as fast as the ship goeth. In like manner they have either a minute of an hour glass, or else a knowne parte of an houre by some number of woordes, or such other lyke, so that the line being vered out, and stopped just with that tyme that the glasse is out, or the number of wordes spoken, which done they hale in the logge or piece of wood again, and looke how many fathomes the shippe hath gone in that time.'

In a later edition of the *Regiment* Bourne gives the warning that the log must 'be so farre astearne that it cometh into quicke water [as opposed to dead water] and the edie of the stearne doth not stay it'. He then adds an example of the calculation as follows: Suppose the time measured is one 120th part of an hour, and the line veered out is 25 fathoms, then the ship goes 120 × 25 fathoms or 3000 fathoms in an hour. An English league is 2500 fathoms, so the ship goes 1$\frac{1}{5}$ leagues an hour. The log should be thrown out every time a change of wind alters the ship's speed, and Bourne explains how to keep the reckoning, adding 'and note it in your Booke', for every ship-master now kept a written record, although it was not yet called a log-book. An incidental remark, too, of Bourne's indicates that the traverse-board (Plate XII) was already in use to peg off the time scaled on each rhumb.

The age-old practice of star-watching at sea had by now been systematized, and Bourne presented a number of star tables, including the position in the Zodiac and declination of thirty-two fixed stars, with the times of their southing at intervals of 15 days, besides a note on each star as to the azimuth of its rising and setting, and the time it was above the horizon. 'The great Dog riseth ESE and setteth WSW and sheweth 9 hours', and so on.

The *Regiment* is avowedly based on Cortes, but there is no slavish following of the Spaniard. Bourne rejects the idea that easting or westing can be measured by the variation of the needle, and declares that 'no maister or pilot of a shippe doth keepe so simple account of the shippes way but that he may know what distance he hath unto any place better than he shall know by the varying of the compass'. He does not, it would seem, know of any 'variation compass' although John Dee had one and taught its use. He merely suggests setting the needle by the Pole Star which he places one-third of a point (3° 45') from the celestial pole, and recommends two instruments only, the Mariners' Ring or astrolabe, and the cross-staff or 'Balla Stella' of which he carefully describes the use. The former was the more convenient when the Sun stands at more than 50°, although the cross-staff was easier to use in a seaway. But the eyes must be protected by dark glass fixed upon the transom, and an error arises because the eye is not truly on the centre line of the instrument. To remedy this he can only suggest 'you must pare away a little of the end of the staff', advice for which he was later to be rebuked by more sophisticated teachers.

The rule for 'raising a degree' introduced a difficulty, for 'your cardes [charts] be most commonly made in Lishborne, in Portugal, in Spayne, or else in France', and on these a distance of $17\frac{1}{2}$ leagues was allowed for every degree of *altura*. This must be altered to the home reckoning of 60 miles of 3 to an English league and Bourne gives the diagram accordingly. He deprecates the elaborate ornamentation of the foreign charts with their illuminated wind-roses and flags, which could (he considered) usefully be substituted by tidal information, or by shore-profiles. The latter were already a feature of the Dutch charts, which were soon to set the standard in England. Bourne was aware of the error introduced into the chart by the parallel north–south rhumbs, and indeed points it out, but on a later page deals with measuring distance on it as though it were scale-true. Whether or no he had seen Mercator's World Chart published a few years earlier, which had been drawn with the express purpose of giving the sailor his parallel rhumbs and yet eliminating his errors, is beside the point. Mercator's map was not fully explained, and therefore was not understood during the author's lifetime.

Bourne's *Regiment* in general, however, is full of good advice. He explains how to plot coastal features from the ship by taking bearings from two points a measured distance apart (Gemma Frisius had long since explained the principle of triangulation), he gives rules for finding roughly the distance of the shore, he explains how to make a simple equinoctial dial of wood which can be tilted to read Sun-time at any latitude. And in a later edition of his book, in which he printed his essay on 'Five Ways to Cathay', he explained how 'a perfect spring clock' could be used to determine the compass points. For one of the 'Ways' led over the Pole, and people objected that it would be impossible to read the Sun since it simply wheeled round the sky at a uniform height. The clock was to have its face marked with twenty-four hours from noon to noon, and as it lay flat each hour would correspond to a direction or rhumb once the hour-hand was pointed to the Sun. So the clock was to be set beside the compass which the helmsman was watching, and he could then set course accordingly. Clocks, of course, were now in common use, and there were clock-makers in London. John Dee had a 'watch-clock', made 'by one Dibbley', which read time to seconds. And in 1555 Richard Eden had printed in his *Decades of the New World* an extract from Gemma Frisius which described how a good clock carried to sea would serve to find the longitude. Bourne did not mention this idea (though he declared that the methods by lunar eclipses and lunar distances were impracticable), but it was present in the mind of William Borough, and within a few years he was trying it out, or rather arranging for someone else to do so.

Borough's mind was always employed upon the improvement of navigation, and as he rose to high office on the Navy Board, his advice was often in request. He interested himself particularly in compass variation, for his far-northern experience had contradicted the common theories about it, and he encouraged the experiments of a clever Ratcliff compass-maker, Robert Norman. Norman had been a seaman, and refers to himself as only an 'unlearned Mechanician', but he had read the available vernacular mathematical textbooks and felt competent to apply them. In the course of his work he had often noticed that after having carefully balanced a needle on its pivot, and then taken it off to magnetize it, it no longer hung true when he put it back again, but dipped down at one end. As, however, he had never heard or read of such an effect of using the stone he took little notice of the matter and simply shortened the down-tipping side of the needle. But on one occasion he was making a very special instrument and in trying to correct the dip he spoiled the long needle. This set him thinking, and he consulted learned and experienced friends, who must certainly have included Borough. He then made the classical experiments which for the first time really established and measured the dip, although the phenomenon itself appears to have been noticed by one or two people at an earlier date. William Borough then encouraged him to write an account of his work and himself wrote a more general tract on the variation, the two tracts being published together in 1581. They were immediately accepted as authoritative, both at home and abroad. Both writers emphasized the dangers resulting from the random use of compasses that had been very variously 'corrected', side by side with charts that had been based on observations perhaps made with a quite different instrument. In the same tract Borough also detailed the historic determination of the variation which he had carried out at Limehouse in 1580, taking nine pairs of forenoon and afternoon shadow azimuths with simultaneous astrolabe readings as Pedro Nuñez advised. His result, $11\frac{1}{4}°$ to $11\frac{1}{2}°$ E., or approximately one point, was the basic figure which later led to the English discovery of secular variation; but it ought to be pointed out that Thomas Digges (Leonard's son) had declared this figure to be the variation in England in a book published in 1571, while it was a value very commonly found on 'compass' sun-dials. Everyone at this date took it for granted that the magnetic variation had a fixed value for each particular place, so that any differences of observation were deemed to be due to errors, and therefore 'influence' cannot be excluded. But Borough's detailed figures appear genuine.

William had become in some ways increasingly conservative as he grew older and he had hard things to say about certain scholars sitting at home who presumed to teach the practical sailor. Perhaps he was thinking of his old teacher John Dee, and of Dee's favourite pupil Thomas Digges, when he declared that for his part he would want nothing better for a journey round the world than an English compass corrected for half a point of variation, and a straight-lined plain chart. For Thomas Digges had been addressing a brief note to the Lord High Admiral on 'Errors in the Art of Navigation', and was preparing a book on the subject. In 1573 Digges had written a tract on the new star that had appeared in Cassiopeia, and had added to it a note on the use of the 10-ft. Radius Astronomicus, with Chancellor's diagonal scale, which was a treasure of John Dee's library. He went on to say that if an unsupported *radius* or cross-staff was used, it should be fitted with sights, or otherwise the observation would be vitiated by parallax. And he then remarked that sailors' observations were so faulty owing to the eccentricity of the eye that loss of life and property was the result. Whereas if pinnules were added to the staff, it would give better results than even the astrolabe which they preferred.

It was in 1576, however, that he wrote of navigation in more general terms, after having apparently read Bourne's *A Regiment for the Sea*. For he tells Lord Clinton that he was determined to deal with errors 'transferred into our language' which sailors took for oracles so that were they not better directed by land-marks and soundings than by their 'Art' there would be many more vessels daily lost than were today. Denying the common view of the relation of magnetic variation to longitude, he de-clared that this was no matter for 'gross mariners' to meddle with. He was surprised at the 'blind boldness of Sebastian Cabot and other such men claiming to have solved the longitude'. It was little wonder that William Borough after reading such language lashed out at the scholar, but Digges had taken the precaution of spending some months at sea, only to be disgusted by sailors' 'gross usage and homely instruments, where half a point com-monly breaks no square'. His list of errors cited begins as might be expected with the use of the plain chart, the next being the false assumption that rhumbs were great circles, whereas they were heliacal lines. The third error was in the method of finding latitude by the Pole Star, due partly (although Digges does not here say so) to the fact that the Star was now less than 3° from the pole and not $3\frac{1}{2}°$ as in the Rule. The fourth error, as he had earlier shown, was due to parallax in the observations taken with the *balestilha* (cross-staff), which was made worse by paring away the end of the instrument to make it fit the eye (the remedy Bourne had suggested). The fifth error lay in the false rule for 'raising a degree' for the course sailed was resolved on the basis of a right-angled (plane) triangle and not the actual spherical triangle. Finally the greatest errors of all resulted from the lack of a method for longitude. To reform all these errors new charts, new instruments and new rules were needed, and in fact for all the defects pointed out, except the longitude, remedies were shortly to be found, although not by Thomas himself as he more than half promised was to be the case.

How alive those concerned were to the need for a new approach to long-distance navigation may be judged from the preparations made for Martin Frobisher's first voyage in search of a north-west passage to Cathay, which took place (after a year's delay) in 1576. He was given an equipment of 'great instruments of brass' more suited to a travelling astronomer than to a seaman who had grown up to such skill as he had in the old, tried methods, by fifteen years or so at sea. Humfrey Cole (now the leading English instrument-maker), and others unnamed, made for him a great blank globe in metal in a case; an *Armilla Ptolomei* or hemispherium; and a *Sphaera Nautica*. None of these three could have been used by an untrained man, nor could the *Annulus Astronomicus*, or astronomer's ring, which cost £1 10s. The *Horologium Universale*, or equinoctial dial, would only be useful on land or in a dead calm, and could only be set correctly when the latitude and compass variation were known. The *Holometrum Geometricum*, a clumsy forerunner of the plane-table, could have been used for land-survey by someone who understood it, and the 'little standing level of brass', perhaps carrying a plummet, would have been handy for setting any instrument. The 'great instrument of brass named *Compassum Meridianum*' costing £4 6s. 8d. was presumably an uncorrected ship's compass of unusual size, and may even have been fitted with a style or with a cross-thread for observations of the variation, since there was no special instrument provided for this. But it is safe to say that the instruments which came into daily use were the 'staffe named Ballestella' made of wood for 13s. 4d., the 'twenty compasses of divers sorts' which altogether cost only three guineas, the eighteen hour-glasses at less than a shilling apiece, and the *Astrolabium* or sea-astrolabe costing three guineas. Very shortly before the ships sailed John Dee came round to Muscovy House in Seething Lane for a conference with Frobisher, who was accompanied by his master, Christopher Hall. Stephen Borough was also there, and the Company's secretary Michale Lok. Lok reported that Dee subsequently took pains to set out the rules of geometry and cosmography for the instruction of the master and mariners in the use of the instruments, and both Frobisher and Hall later together sent him a letter of thanks. But its tone suggests that the thanks were rather for his good intentions than for any success in his teaching. They had been supplied also with a 'very great chart of navigation' which cost £5, probably a world chart such as was made by the Dieppe school of chart-makers. In addition they had a printed copy of Mercator's 'Hydrographic' world map of 1569, which quite apart from its conjectural outlines of the Arctic regions was on the author's new projection, which is not useful in high latitudes. Of the half-dozen sheets of parchment ruled for plotting the new discoveries, six are reported to have been 'ruled plain', that is to say with the usual pattern of rhumb lines, while two were ruled 'round', which may possibly mean with latitudes and longitudes on a zenithal projection and the heliacal rhumb lines to the use of which William Borough elsewhere makes some reference. A copy of Medina's *Arte de Navegar* in Spanish was provided, but no doubt Christopher Hall, who was a very skilled man for his day, possessed his own copy of Eden's Cortes or Bourne's *Regiment*.

Francis Drake, when he went off quietly 'for Alexandria' in the following year, provided for his navigation by always kidnapping a pilot who knew the particular part of his route, and to decide his general plans he bought a large world chart in Lisbon. This was probably the work of a very notable Portuguese chartmaker, Vaz Dourado, which would have given him excellent detail of Magellan's Strait and the East Indies. Foreign pilots always carried their own set of charts, sailing directions and instruments, and when Drake captured a ship it was his habit to seize these and sometimes throw them overboard. Like Frobisher he was a professional sailor ('This Corsair . . . so well versed in all modes of Navigation', said Pedro Sarmiento), but it was still more usual to have a man of rank, necessarily therefore a military man, in chief command—as for example Sir Hugh Willoughby and Sir Humfrey Gilbert.

Gilbert had studied the subject of exploration for nearly twenty years. He had put himself to school with men like the elder Richard Hakluyt and John Dee, and he took with him in 1583 technicians, properly instructed and equipped. But he used his authority to override the master's judgment, and so the *Delight* was cast away. Not only was the ship lost but 'the cardes [charts] and plats [plans] that were drawing, with the due gradation of the harbours, bayes and capes, did perish with our Admirall'. Remembering his losses, Gilbert used to fall into a rage and beat his unfortunate boy, but he was soon to be lost himself, dying with a noble saying on his lips. The instructions survive which were prepared for his professional land-surveyor in the year 1582, and their particular interest for the history of navigation lies in the fact that the charts and maps were to show latitude and longitude, so that (as Edward Hayes later reported) 'some observed the elevation of the pole, and drew plats of the country *exactly graded*' when they were ashore in St. John's. The surveyor, Thomas Bavin, and his assistants were to carry the instruments advised in his tract of 1581 by William Borough, including those for the dip and the variation of the needle, as well as three flat 'watch-clocks' reading to minutes. The latter were for the longitude, since they could (it was supposed) be kept wound up and carry English time which could be compared with that found by the ship's equinoctial or universal dial. Alternatively, since a solar and a lunar eclipse were foretold for 1582, one of the clocks, set by the dial, could be used for timing the eclipse. Actually the expedition did not sail that year, but the references to chart-making already quoted make it clear that the same general programme was being carried out in 1583.

A very few years later a brilliant mathematician, Thomas Hariot, was to turn the light of his genius upon the problems of navigation, but meanwhile a Dutch mariner had produced a combined nautical manual and chart book which was to become an essential part of every sea-master's equipment under the title of a 'Waggoner'. This publication gives a very clear picture of the degree of skill that had been reached, and the standards attained by the sailors who fought the Spanish Armada, and it marks in particular the opening of a period during which the Dutch were reckoned as the masters of the seaman's art. Dutch chart-making can be traced back to about 1527, when an engraver and painter, Cornelis Anthonisz, settled in Amsterdam and during the next thirty years produced rutters, maps and charts. Like Gemma Frisius and Gerard Mercator he reaped the advantage of the Imperial Court being kept in the Low Countries, and in fact accompanied the Emperor's expedition to Tunis in 1541. He was a contributor to the *Zeekaartboek* of Wisby published in 1551, and the forerunner of Lucas Waghenaer who gave his name to the Waggoner.

Lucas was a seaman who returned to a shore job which he lost through some financial irregularity. So he turned his mind to compiling a new sea manual, with the advantage that there were engravers of outstanding quality among his fellow-countrymen. The title-page of the *Spieghel der Zeevaerdt*, 1584 (Englished as *The Mariner's Mirror* 1588, Plate I), was engraved by Johannes à Doetecum, and all the current charts and harbour plans were reduced to a uniform and moderate page size and reproduced in uniform style, with a key to the symbols used for depths (at half-flood and half-ebb), channels, anchorages and dangers. On the back of each chart were sailing directions and a certain amount of commercial information about the particular cities or countries on it. The preliminary text began, as the Spanish manuals did, with tables and directions for finding the Epact and Golden Number. This was followed by a table of New Moons throughout the months of the nineteen-year cycle from 1585, and a new table of solar declination for 1585–8 drawn up by a 'geometer and State engineer', both with directions for use and worked examples. Next there was a catalogue of declinations and right ascensions of 100 fixed stars, with instructions as to their use to find the time and the height of the pole, besides a shorter list of six stars for easy use with the cross-staff. For example: Alhabor was in the meridian when the Guards were aloft in the south-east; Azimech was in the meridian when the Guards were north and by east. There were the usual Rules and diagrams for the North Star (based on $3\frac{1}{2}°$ from the pole), for 'shifting the tides', and for raising a degree, reckoned at 15 Dutch leagues. In addition there was a table of the Sun's right ascension through the year, another of a large number of coastal bearings and distances, and a section on the method of marking the scale on the 'Graetbogen' or cross-staff.

In his exhortation to the young apprentice to the sea, Waghenaer tells him that the astrolabe and cross-staff, next to the mariners' compass, are the chief instruments he must learn to use, but first of all he must practise taking soundings and drawing landmarks. *The Mariner's Mirror* was a resounding success. Lord Charles Howard admired it, and Sir Christopher Hatton had it translated into English with the charts re-engraved in London. But it was already out of date. Its author knew nothing of the 'Errors of Navigation' pointed out by Thomas Digges. And in France Michel Coignet, writing on the same subject in 1581, had stated that the common table for raising a degree, being based on straight-lined rhumbs, was increasingly inaccurate in latitudes beyond 50°, where the rhumb distances are so much longer than the great-circle distances between any two points. For the fourth, fifth, sixth and seventh rhumbs he gave the following pairs of figures, showing the distances (in leagues) necessary to raise a degree at the equator and at Lat. 60° respectively.

Rhumb	Lat. 0°	Lat. 60°
4	$24\frac{19}{24}$	$25\frac{1}{12}$
5	$31\frac{1}{2}$	$32\frac{1}{12}$
6	$45\frac{9}{24}$	$47\frac{1}{4}$
7	$89\frac{5}{6}$	$100\frac{1}{3}$

The time was in fact at hand for the general correction of the crude figures and rough observations that had hitherto been 'near enough' for use at sea. In particular, a knowledge of trigonometry and of trigonometrical tables was penetrating beyond the narrow circle of the astronomers as the more enquiring seamen began to look into books of Ephemerides for themselves instead of being satisfied with the extracts found in manuals. And while the enquiring seaman himself might be a rarity, there was a growing body of teachers of navigation, many of them university men, who were concerned to advance the subject.

Yet the earliest Englishman to open in print the question of using mathematical tables was the self-educated seaman, William Borough, writing his tract of 1581. A reader of the collected works of Pedro Nuñez, he knew how to calculate the Sun's azimuth instead of observing it, when it was required for a reading of the compass variation.

'If the Reader is delighted with variety of demonstration of this matter [he writes] let him peruse the 14 proposition of the 4 [chapter] of Regiomontanus, and the 13 proposition of the 14 chapter of the first book of Copernicus. . . . In these examples I have used ye abridged table of 100,000 the whole sine, which though it give some ease in the working, yet it is not so exact as that of 10,000,000 of Erasmus Reinholdus. Unto the which, with his Canon Secundus answerable to the same, if the third Canon of the Hypoteneuses [secants] were annexed, we should have an entire table for the Doctrine of Triangles, that might worthily be called the Table of Tables. Whiche thyng though Georgius Joachimus Rheticus have well begun, and framed it orderly from ten minutes to ten, yet it is left very rawly for such as desire the exact truth of thynges. I have therefore for myne own ease and use calculated the complement of this table, and almost ended it for the whole quadrant from minute to minute; which if in the meane time before I have finished I shall not find it extant by any other I will publish it for the commoditie of all such as shall have occasion to use the same for Navigation and Cosmographie.'

There was in fact no need for Borough to labour over the figures. The massive and complete tables of Rheticus for ten-second intervals were published in 1596, while the straightforward textbooks of trigonometry of Christopher Clavius (1586) and Bartholomew Pitiscus played their part in ushering in the new period of the approach to mathematical navigation.

XVI. A seventeenth-century maritime chart by Nicholas Comberford of Ratcliff (*fl*. 1646–1666). Although printed charts had come into use, naval captains often preferred them specially drawn.

XVII. An Italian Mariner's Compass dated 1719.

PART FIVE

TOWARDS MATHE-MATICAL NAVIGATION

X

The True Chart

1. THE NAUTICAL TRIANGLE

As the sixteenth century drew towards its close, advances were being made in pure mathematics and astronomy which could not fail to have their repercussions in the maritime world. If not the sailors themselves, yet their teachers were making use of the new Ephemerides of Stadius, Maestlin and Evaerts; they were becoming practised in the new trigonometry, and were realizing from the more refined observations of men like Tycho Brahe the gross inaccuracies of the old Alphonsine Tables and derivative calendars. It was no longer possible to believe, as Jean Rotz had done, that the Moon moved at a uniform 30' along her orbit every hour, while it had been pointed out that an error of as little as 5' in reading her position would produce an error of 2° 5' in calculating the longitude. And with sailors' instruments it was hard to read even to half a degree; at very best they might approximate to 15'. In Spain and Portugal navigation to the Indies had settled down into a fixed routine, but in Elizabethan England men's minds were full of projects for colonization, discovery and naval warfare. And some of them at least realized that the necessary improvement of methods of navigation was at base a mathematician's problem. Sir Walter Ralegh took Thomas Hariot, a young Oxford scholar, into his service, while a few years later the Earl of Cumberland engaged Edward Wright, a Fellow of Caius College, Cambridge, for the same specific purpose, to study the problems of navigation. There were others besides (like Lord Lumley) who sought out and encouraged scholars. Both Hariot and Wright were familiar with Thomas Digges's summary of the current Errors of Navigation, and both confirmed them for themselves on the single voyages they made—Hariot to Virginia in 1585, Wright to the Azores in 1589.

On these voyages the two mathematicians must have learned

how wide a gap there lay between themselves and the practising sea-masters, a difference not only in mathematical knowledge and capacity, but in habit of mind. For the sailor worked by rule and by tradition, so that he was always likely to think the old way best. Yet we have only to study the Journal of any voyage of the day to realize the perils into which his imperfect methods led him. Captain Luke Ward, for example, relates his voyage of 1582 made on the sister ship to Captain Edward Fenton's *Gallion Leicester*. In mid-Atlantic Fenton· called a conference of officers which included Christopher Hall, who had been with Frobisher in the Arctic, and Thomas Blacoller, one of Drake's old pilots. 'Divers of their charts and reckonings were shewed', says Ward: 'by some it appeared we were 115 leagues, by some 140 leagues, and some a great deal further short of Brazil. But all agreed to be within 20' of the line, some to the north, some to the south.' They differed, in short, by nearly a hundred miles in their east-westing, and by as much as forty miles in their latitude. They were in equally bad case in May 1583 when they were nearing home again. They had sighted land 5 leagues off to the north-east, and according to practice took soundings and examined the bottom material. But some said they were off Ushant, others that the landfall was the headland of Fontenay (Pointe du Raz). Darkness was falling, and so, again following the usual practice, they went about and 'lay south-south-west six glasses, and then went about and lay north-west six glasses'. These were half-hour sand-glasses, and in the six hours the May night was over. Actually they found they were off the Scilly Isles, and what they had taken for the 'Seames' (Sein) turned out to be (so Ward says) the Bishop and his Clerks. In other words they were nearly two degrees of latitude, and about 100 miles of longitude out in their reckoning.

Equally hair-raising was the experience of a home-coming fleet under the Earl of Cumberland in 1597, described by an unnamed gentleman aboard, possibly William Monson.

'In the evening while yet it was fair his lordship commanded the lead should be wet, and at the second sounding, partly by the depth of the water and partly by the ground it was reasonably judged we were nearly entering into the sleeve [the Channel]. Marry, whether to the coast of France or our own coast there was difference of opinion. For the ground was like a mealy bran and the depth, as I remember, fifty-five beside allowance for the stray. But among the branage sand there was found a number of little black corns, upon the likeness by seamen called pop-corns. And these to be found upon our coast or upon the coast of Scilly, it was very constantly denied by some who had beaten the Channel very much and often: yet our master and the most with him made us upon the bank of Scilly, and his Honour was content to let his opinion prevail, the rather that the weather was yet fair and the wind large to lay it off or on upon any occasion.'

A storm, however, sprang up and after an interval

'the ship being laid by the lee, they sounded again, and found so much alike in the same depth, but the ground upon the tallow, did still more and more assure us of being in the Sleeve: only the scallop shells confirmed their opinion who held us rather upon the coast of France. Our master would still have held the same course, north-east and by east, but his lordship about midnight absolutely commanded otherwise and gave instruction to sail a more northerly course, which the event showed was the saving of us all from utmost danger. For the next morning very early we saw land and quickly it was made Normandy, so that clear it is that when we began to alter our course we were exceeding near Ushant and the rocks, upon which if we had fallen in the night, there had been very little twixt us and sudden death.'

Little wonder that William Bourne in his *Regiment* (still the most widely used English manual) had laid it down as the first qualification of the man 'Meetest to take charge', that he should be a good coaster. This meant that he must know the Moon's course, 'whereby he doth know at what time it is full Sea or a low Water, knowing in what Quarter or part of the Sky that the Moon doth make a full Sea at that place. . . . He ought to be expert how the tide gates or currents do set from place to place.' And above all he must know the soundings and the grounds and every place 'by the sight thereof'. Only if he is to take long voyages need he also have knowledge of plats and cards (charts), understand the instruments for taking the Sun or the star, and be able to correct his instruments, and handle the calculations for working out positions by the Sun.

Richard Polter, an old sea-master who in 1605 held the important office of Master of the King's Ships, affords an excellent example of the type of man with whom the reformers of navigation found it so difficult to deal. There is much sound common sense in his *Pathway to Perfect Sayling*, but also much ignorance and obstinacy. He considered sailing under six heads: Chart, Compass,

Tide, Time, Wind, Way, and he is obsessed throughout by the apparent anomaly that a great circle passing through the zenith and nadir of a point cuts its meridian at right angles, and yet the east–west rhumb (which by definition also cuts the north–south meridian at right angles) is said to be a small circle. He is unable to grasp, too, the idea of a negligible error, and therefore rejects entirely the method of finding the meridian by equal altitudes because the solar declination will have changed between the readings, although the change is at most by only one minute an hour. He has discovered, too, from the Ephemerides that the Sun is not a perfect timekeeper, still less is the Moon regular in her motions, and is aware that a mistake of nearly an hour and three-quarters is possible in calculating the time of high water by the usual rules. The 'running glass' he considers the safest timekeeper for plotting the course, and indeed it was the usual practice to peg the traverse-board in terms of half-hour glasses. Polter's warnings therefore about the errors to which a worn or badly made glass is subject were timely. But he goes too far again over the magnetic compass, declaring that no two lodestones are alike, and that the needles therefore do not necessarily stand alike. Hence 'there cannot generally be set down a certain variation for any one place'. A few years earlier Edward Wright had dedicated to him his translation of Simon Stevin's *Haven-finding Art*, in which there was a list of variations at particular places, intended to be used in combination with their latitude for identifying these positions. Evidently Polter would have none of it. Indeed, he does not even accept the use of log and line for finding way.

Polter's *Pathway* was put on sale again after an interval of nearly forty years, so that it may be judged how far the general rank and file of seamen were from profiting by the work of such men as Hariot and Edward Wright. The first-named did not, indeed, publish what he wrote, which was intended for Ralegh and his ship's officers, but his ideas spread nevertheless among teachers of navigation.

In his voyage to Virginia Hariot had tested the seamen's methods of observation, and noted particularly their complaint that the latitude obtained by observation of the star differed by nearly as much as a degree from that found by the Sun. This he rightly put down to the faulty Regiment of the North Star, and he made continuous observations on the true position of the Pole Star (α–Polaris) with a 12-ft. instrument (probably an Astronomer's Staff) which he set up on the roof of Durham House, Sir Walter Ralegh's mansion in the Strand. These he compared with an observation made with a similar large instrument by Gemma Frisius at Louvain in 1547, which had given a polar distance of 3° 8′. Hariot concluded that, owing to precession of the equinoxes, the figure altered by about 24′ in a century, and that in 1598 the true value would be 2° 55′. He further pointed out that the corrections to be made, according to the eight positions of the Guards, to obtain the true height of the Pole from the star, depended upon the latitude of the observer. Both Pedro Nuñez and Thomas Digges (he said) were aware of this point, but could not explain how to deal with it. He himself therefore prepared a correction table for every tenth degree of latitude (10°–60°) with a difference column for use in interpolation. He further drew up a new table of solar declinations serving for the two four-year periods 1593–6, 1597–1600, dating from the beginning of Ralegh's Guiana voyages. Here, too, there was a difference column, since, as Hariot pointed out, the time change between local noon and the noon of the table became significant once the ship was a certain distance (he put it at an hour angle of 30′) east or west. Such a refinement, involving corrections of a few minutes only, was in a sense absurd, owing to the fact that much greater errors arose from the imperfections of the instruments used, and from the difficulty of observation at sea. The same objection might also be urged against Hariot's introduction of correction for the dip of the horizon, for according to his table (which is consistently about a minute high) a reading even from the poop would be subject only to a correction (subtraction) of seven or eight minutes. He himself agreed that corrections for the Sun's horizontal parallax and for atmospheric refraction could be omitted, although he dealt with them in his teaching manual *Arcticon*, a manuscript work unfortunately lost. On the other hand, all these corrections had eventually to be applied, and it was important that they should be brought to the notice of seamen and instrument-makers. Both a table of the dip of the horizon, continued up to an eye level of 90 ft. (dip 11'), and a table of Sun's parallax (using the current maximum of nearly 3', which was much too high) were printed by Edward Wright in 1599. Following Kepler he rightly considered stellar parallax negligible, as did Hariot.

Hariot took up carefully the question of instrumental error, and in particular that due to parallax of the eye in using the cross-staff, since this was the instrument that seamen preferred before all others. The correction that he found necessary for the eccentricity of the eye amounted to over $1° 30'$, although it varied with the setting of the eyes between individuals. When the staff was used to sight the Sun he advised that the upper limb of that body should be taken, the transom lying across it and so protecting the eye: the pieces of dark glass which were commonly attached only led to distortion. Such was his advice, and it involved reducing each reading by 16', the accepted semi-diameter of the Sun. An alternative method of avoiding glare was to use a back-staff, and in Hariot's papers there are several designs for such an instrument. John Davis' back-staff or quadrant, which became a standard instrument for over a century (Plate XIII), was (as he himself said when first describing it) an improved pattern of a staff in which the general principle of observing the shadow cast by the Sun was already in use.

With regard to swinging instruments, such as the sea-astrolabe and the sea-ring, Hariot pointed out that a true reading could be taken by finding the mean point between the extreme readings. He concerned himself, too, with the variation of the compass, and, by drawing up a table of the Sun's amplitude for each whole degree of declination and each whole degree of latitude up to 54° (giving difference columns for interpolation), he made it possible for sailors to correct the compass daily at sunrise or sunset. Again, of course, his table was not published, it was for Ralegh's use, and the earliest printed table of amplitudes is dated 1608. This is found in the *Hydrografia* of the Portuguese pilot Manuel de Figueiredo, which was shortly afterwards translated and the table extended by Nicholas Lebon of Dieppe, and extended once more (to Lat. 66°) by his successor Jean Le Tellier. The latter (an East India captain) published the table and the method of its use in 1631. No English tables appeared until 1664.

XVIII. The seaman's astrolabe was made very heavy, with a strong swivel thumb-ring. The instrument pictured was recovered from the sea in 1845, and may have come from a wrecked vessel of the Spanish Armada.

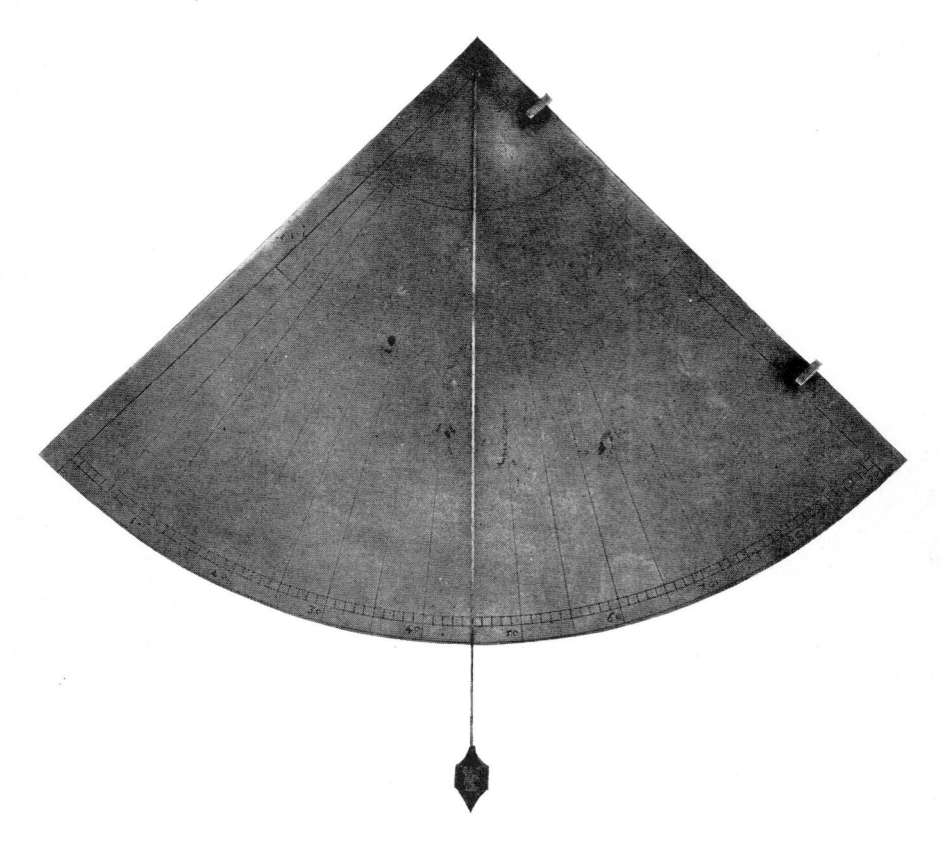

XIX. The seaman's quadrant indicated the altitude of a heavenly body viewed through the sight holes by means of a plumb line, here falling across the scale at 45°.

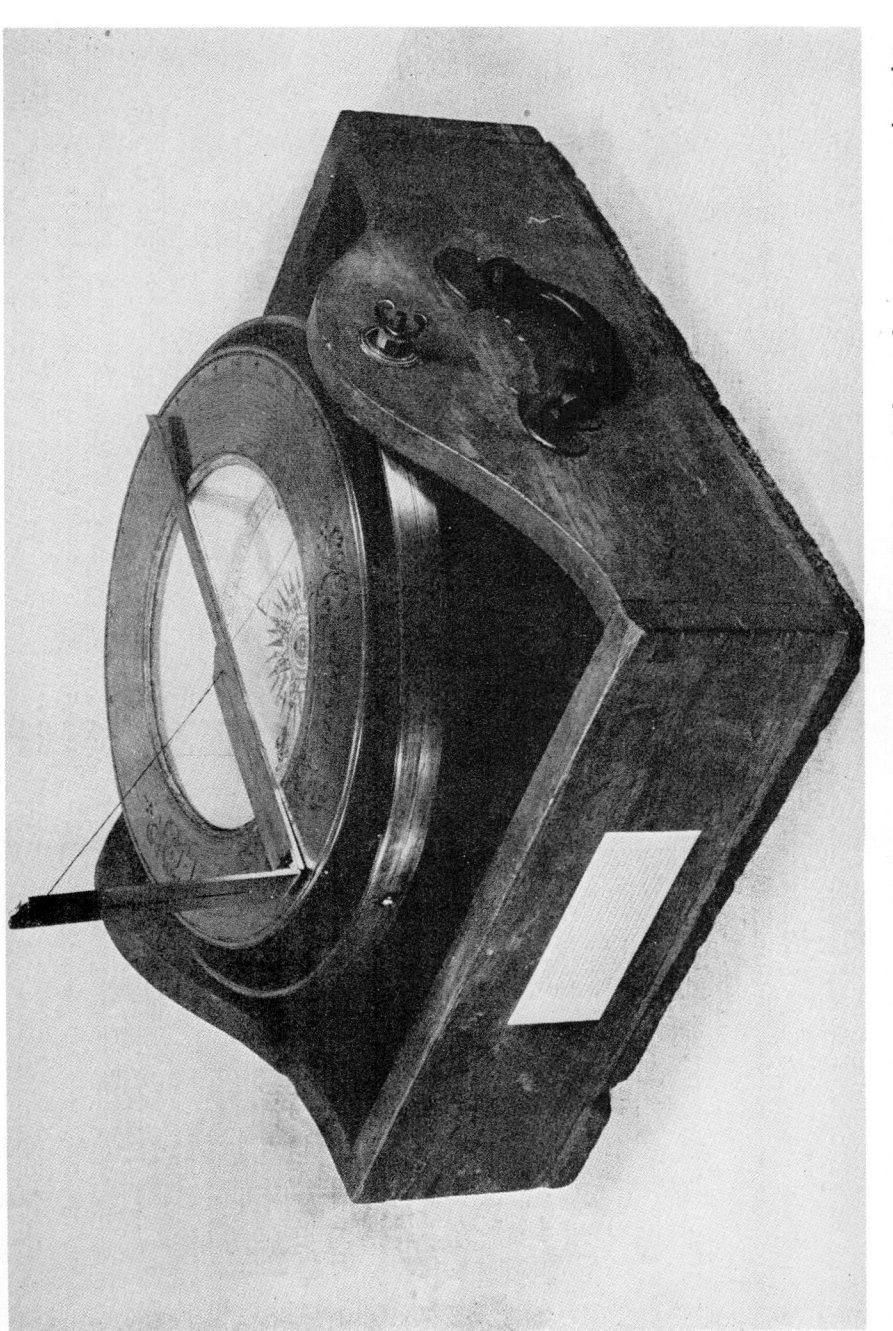

XX. The azimuth compass has a brass rim fitted across the compass box to which is attached an index carrying an upright sighting slit and thread. By means of marks within the box the instrument is set to the compass card, and the index is turned until the observer sees through the slit that the shadow cast by the thread coincides with the fiducial line.

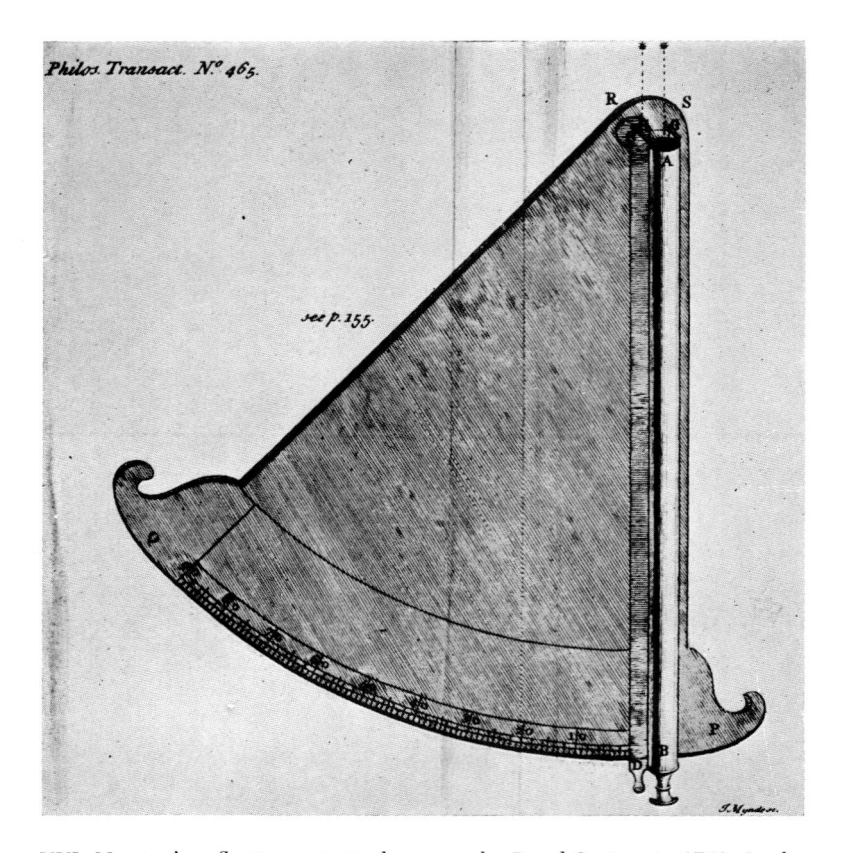

see p. 155.

XXI. Newton's reflecting octant, shown to the Royal Society in 1742. In the zero position, here shown, the same star is seen directly and by reflection between the two parallel mirrors through the tube AB. By moving the arm CD the image of a second heavenly body can be sent by double reflection down the tube. The actual angle moved is half that between the two bodies, hence the scale readings must be doubled. (From *Philosophical Transactions of the Royal Society*, No. 465, 1744.)

The example of working out a variation problem which Hariot offered was as follows:

10 Feb. 1595. You are south-west of the Lizard in a height of 48°.
The sun rises at 7° S. of E. by the meridian compass.
The Book of the Sun's Regiment gives the declination at noon 10° 58′ S.
On the day before it was 21′ more, so that at sunrise on Feb. 10 it is $\frac{21}{4}$′ more than at noon, or 5′ more.
Hence the correct figure is 10° 58′ + 5′ or 11° 3′, but the figure 11° is near enough.
The Table of Amplitudes for Lat. 48° decl. 11° gives 16° 33′. Hence the variation is $9\frac{1}{2}$° E. of N.

As an example of latitude by the Pole Star he gives:

An other example or praesident upon an East guard.

Apparent altitude of the starre by the lesse staffe	$\overline{4}2.30′$
Parallaxis of the staffe $\overline{0}.32′$ $\Big\}$ to be abated	$\overline{0}.37′$
Surplus of the Horizon $\overline{0}.5′$	
Therefore the true altitude of the starre	$\overline{4}1.53′$
The allowance against $\overline{4}0$ because it is next under an East guard to be added	$\overline{1}.39′$
Therefore the altitude of the Pole	$\overline{4}3.32′$

By that altitude of the Pole seek a precise allowance.
Against $\overline{4}0$ you have the same allowance as before $\overline{1}.39′$. The difference for the next underneath is 6 minutes which multiply by the odd $3.32′$ or $\overline{4}$ because the minutes are more than half a degree and the summe will be 24 which hath two tennes and therefore 2 minutes are to be added to the allowance above | $0.2′$

Therefore the allowance is	$\overline{1}.41′$
Which add as by the title to the true altitude of the starre	
Then the precise altitude of the starre is	$\overline{4}3.34′$

And for the latitude by the Sun he works as follows:

Meridian alt. of the higher edge of the sun by the back-staff		$\overline{7}9.35′$
Parallaxis of the staffe $\overline{1}.35′$ $\Big\}$ to be abated		
Surplus of the horizon $\overline{0}.5′$		$1.56′$
Semi-diameter of the Sun $\overline{0}.16′$		
True meridian alt. of same is		$\overline{7}7.39′$
Declination in the Regiment of the Sun	$\overline{2}.56′$	
Part proportional to be added for 900 leagues	$\overline{0}.3′$	
True declination northerly		$\overline{2}.59′$
Abate because zenith is more northerly		$\overline{7}4.40′$
Altitude from equinoctial		$\overline{1}5.20′$

So far Thomas Hariot was merely attempting to improve the working of the current methods of navigation. Of greater importance was the problem of the plain chart, by which, as every critic agreed, a correct course could not be laid down. Mercator's map devised 'for sailors' was of course known, but William Borough (for example) found its unequal spacing of the lines of latitude intolerable. 'In our charts for S. Nicholas and Narve' (he says), the coasts are described by the common sailing compass 'with consideration of the variations at divers places, whereby the true meridians, reformedly set down [i.e. correctly spaced at mid-latitude] . . . do necessarily widen northwards and straighten [i.e. narrow] southwards, contrary to the true form and nature of meridians. And yet notwithstanding that is the best means hitherto known to reform in plat the errors that else would grow by the strange variations that way.' Ten years later (in 1592) a Cambridge scholar, Thomas Hood, who had taken up the profession of teacher of navigation in London, stated that he had written a book 'concerning the use of Mercator's Card', but had not leisure to explain it to his pupils. This excuse arouses a strong suspicion that he did not feel equal to the task.

Mercator's production was, it must be admitted, essentially a scholar's map. It was not drawn according to the conventions of a sea-chart, and it included conjectural coastlines based on literary sources. The legends were in Latin and the principle of the new projection was not explained nor its method of construction indicated. True there was an inset diagram and a partial explanation of its use for the graphical solution of a nautical triangle, but the ordinary sailor could have made nothing of it. To the mathematician the general idea involved would soon become clear (Fig. 27). On the plain chart the distances between the lines of latitude are correctly spaced, while those between the lines of longitude (kept parallel) are progressively increased in the proportion of the secant of the latitude. For example, the distance between the meridians at latitude $60°$ should be half that at the equator. On the plain chart it is equal to that at the equator, that is to say it is twice the correct distance (secant $60° = 2$). For the sailor the consequence was that the proportional relation of northing (latitude) to easting (longitude) was everywhere falsified, and so any course laid was incorrect. Mercator's remedy was to falsify the spacing of the lines of latitude in exactly the same proportion as that of the plain chart's parallel lines of longitude. This would make all angles (i.e. rhumbs) correct, and courses laid consequently true, while in any small area the scale in all directions would be the same. But this scale would differ from latitude to latitude, increasing progressively according to the table of secants. By 1569 Mercator had at his disposal this necessary trigonometrical table to draw his map, although it is not clear exactly how he used it. He could have stepped off his central scale of latitudes from the equator at distances $x \sec 1°$, $x \sec 2°$, $x \sec 3°$ . . . where x was the length he assigned to a degree of the equator. Any such procedure would mean that the scale increased in jerks at each degree, but on the relatively small scale of his world map this would be immaterial, although it is not mathematically acceptable, as the true increase of distance between the meridians is by infinitesimals.

It is clear, however, that both Thomas Hariot and Edward Wright at some date before 1590 were thinking afresh about this method of drawing a chart on which the rhumbs would be correctly shown by straight lines drawn from point to point by the pilot's ruler. Both men constructed the tables of meridional parts (i.e. spacing of the lines of latitude) for their charts by a continuous addition of secants at 1′ intervals, Hariot using the trigonometrical tables published by Christopher Clavius in 1586, and Wright those of G. J. Rheticus, published in 1596, at least for the final version of his figures. These were printed for ten-minute intervals in Wright's famous work, *Certaine Errors in Navigation*, 1599, and in their full form for one-minute intervals in the later edition of 1610. Hariot did not publish his *Canon*, but from what may be considered a first manuscript draft of his work on the subject, entitled *The Doctrine of Nauticall Triangles Compendious*, it is clear that he had begun the calculation of meridional parts (M) by the equivalent of the modern formula $M = K \log \tan (45° + \frac{1}{2}\phi)$ although the integral calculus had not yet been introduced or logarithms invented. This, however, while a tribute to his mathematical genius, is irrelevant to the history of navigation, nor has the actual map he refers to as drawn on the projection survived, unless indeed it was the basis of the world map added to the 1599–1600 issue of his friend Hakluyt's *Voyages*. Edward Wright does not claim this map, and on the large world map which he published in 1610 all he says is: 'So far as I can yet learn I first published to the world [in 1599] the exact way to make the parts of the meridians and parallels keep the correct proportions.' At this earlier date he drew a map by means of his table (an addition of secants) to illustrate the journey he had made to the Azores (the Earl of Cumberland's voyage) some ten years earlier.

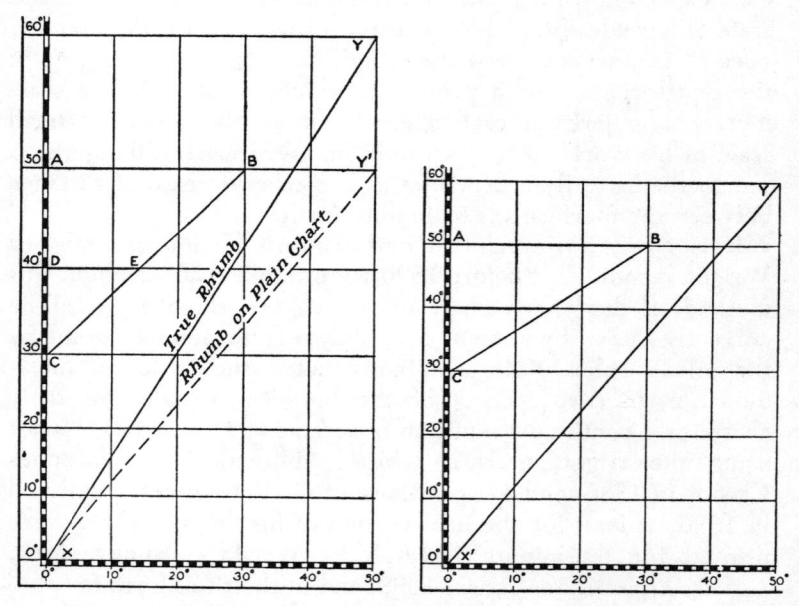

Fig. 27. A rhumb line (XY) and nautical triangle (ABC) on Mercator's projection and the plain chart.

On the Wright-Mercator projection the sailor could draw for the first time a 'nautical triangle' which showed latitude and longitude, direction and course correctly. It was a right-angled triangle, and hence if two sides were known, or one side and another angle the whole triangle could be solved by trigonometry (Mercator had been obliged to offer only a graphical solution). But the difficulty was the change of scale from one end of the hypotenuse of the triangle to the other, i.e. it lay in measuring the length of the course. This is met by using the scale on the meridian midway between the places of departure and arrival, which became known as 'mid-latitude sailing'. It is only approximately correct, since the meridian scale does not change by equal increments on either side of the mid-latitude point; consequently with the more refined methods of modern sailing a further correction has to be made (Fig. 27). Hariot, although he set out the various solutions of the nautical triangle according to the particular elements that were known, did not deal with the matter of scale in his draft, while all that was said on the map of 1600 was: 'If [the two places] differ in latitude see how many degrees of the meridian taken about the midst of that difference are contayned between them and so many leagues is the difference.' The user was expected, that is to say, to step off the distance with his dividers opened to the width of the mid-degree as shown on the marginal scale of degrees. Each of these represented 20 leagues according to the English reckoning.

Thomas Hariot appears to have ceased to interest himself in navigation after Ralegh fell into disgrace, or perhaps he considered he had solved all the problems of navigation that were capable of solution, for among his papers is a little jingle which he entitled 'Three Sea Marriages' which runs as follows:

> 'Three new marriages now are made
> One of the staff and astrolabe.
> Of the sun and star is another,
> Which now agree like sister and brother.
> And card and compass which were at bate
> Will now agree like master and mate.'

He had, in fact, dealt with contradictory instruments, supplied sailors with more accurate tables and prepared a 'true' chart.

The introduction of the 'true' chart, which any chart-maker could now draw with the help of Wright's tables, was the most important advance in navigational technique since the Portuguese astronomers had first taught the use of solar declination. And it coincided with the introduction of trigonometry into general use. A translation of Pitiscus's textbook (1600) on the latter subject was made for his pupils by a London teacher of mathematics and navigation, Raphe Handson, and he applied it to a demonstration of the superiority of the Mercator over the plain chart, explaining clearly the method of 'mid-latitude' sailing. Richard Hakluyt persuaded him to publish his translation and its addenda for general use, and this Handson did in 1614. Meanwhile Edward Wright himself was engaged to deliver a regular lecture on navigation on behalf of the East India Company which unfortunately was not renewed after his death in 1616. He had then almost completed a translation of Napier's 'Admirable Table of Logarithms' as he termed it, and this new help to accurate calculation was immediately taken up by practical teachers, and when not long afterwards tables of natural logarithms and of logarithms of the trigonometrical functions were compiled, they added these to their manuals.

Even before the date of the appearance of Wright's *Certaine Errors*, it was becoming generally realized by seamen that more up-to-date tables were necessary than those in Cortes's manual and its derivatives, and a teacher of navigation, John Tapp, who had entered business as a bookseller and stationer, decided to meet the need. He issued in 1601 a *Seaman's Kalendar, or an Ephemerides of the Sun, Moon and most notable fixed Stars*, which continued to appear throughout the century under different editors, and may be considered the prototype of the *Nautical Almanac* although its preparation was never in the hands of academic mathematicians. Indeed it was written in homely style suited to poorly educated readers, and contained instructions on how to use the various tables, as well as rules for working out the course made good from the plain chart. The use of a traverse board with graduated edges (after the fashion of Gemma Frisius's geometrical square) was even suggested for those with little arithmetic. Tycho Brahe had published a new catalogue of the fixed stars in 1600, and this was drawn upon, while it was now becoming possible to build up an increasingly long table of astronomically determined longitudes. The day-to-day tables gave the Sun's right ascension and declination, and the day and hour of the New Moons, while there was a detailed explanation of how to 'shift the tides'.

To his new edition of William Borough's *Discourse of the Variation* (1614) John Tapp also added a table of sines, tangents and secants to a radix of 100,000, and posed the six questions on the nautical triangle in what had by now become the standard way:

1. Given Difference of latitude (D. lat.) and course, what is the distance?
2. Given D. lat. and course, what is the departure from the Meridian? (i.e. D. long.)
3. Given D. lat. and distance what is the course?
4. Given D. lat. and distance what is the departure?
5. Given D. lat. and departure what is the course?
6. Given D. lat. and departure what is the distance?

In the older manuals figures for D. lat. and departure had only been given in respect of each whole point of the compass, but Tapp now suggested that working ought to be to half-points, and in fact very shortly complete tables of distance and departure for each degree from the meridian were to appear.

Side by side with the improvement of tables went the improvement of nautical instruments. Many of the new designs were too elaborate for sea use, such as Edward Wright's sea-rings, a set of astronomer's rings fixed above a large mariners' compass, but an azimuth compass, the rim of the compass box graduated in degrees and carrying an alidade, became a standard instrument for measuring the variation. Much more important, however, was the development of the telescope, although it was long before it could be actually handled by sailors. Hariot showed a perspective glass to the Indians of Virginia in 1585, and in his will he left to his patron the Earl of Northumberland 'my two perspective trunckes wherein I use espetially to see Venus horned like the Moone and spots on the Sun'. According to Dr. Mark Ridley, Edward Wright (who had a great reputation also as an instrument-maker) actually attached perspectives to his instruments, perhaps merely as a 'finder' in his survey work, but the true value of the telescope lay in the first instance in the advances which it made possible in astronomy. Meanwhile in 1609 Kepler had published his *Astronomia Nova* demonstrating the elliptical orbits of the planets, and in 1627 produced the Rudolphine Tables (Ephemerides) which became the standard for the seventeenth century.

Of more immediate value to sailors faced with increasing demands for mathematical competence, was (in England at least) the interest in the problems shown by the early Gresham College

Professors. Henry Briggs, Gresham professor of geometry, had given considerable assistance to Edward Wright, and Edmund Gunter, the third professor of astronomy at Gresham College, gave definitive teaching on navigational problems, and dealt in his textbooks with their solution, besides devising instruments for sailors' use. In this last respect he was greatly assisted by the fact that among English instrument-makers there were some who had developed a very high degree of accuracy in dividing and engraving scale graduations. Especially notable were John Thompson, who worked in wood, and Elias Allen, who worked in metal. Gunter's sector (which folded like a carpenter's rule) carried on one side scales of sines, secants and tangents, and on the other a scale of meridional parts, besides three or four scales for other purposes. An important one of these, which went all along one edge of the opened-out sector, was the Line of Lines, a set of equal divisions, subdivided into tenths, to which any desired value or unit might be assigned. An example of its use will be given presently. Meanwhile it should be mentioned that Gunter also gave a table of meridional parts for every tenth of a degree, and a table of D. long. and distance resulting from sailing any D. lat. from 1° to 90° on each of the rhumbs. He explains in detail how to draw the network of a chart for an area between Lats. 50° and 55°, and on this works out his solutions of the nautical triangle, besides giving a second working from the sector.. He so chooses his examples that the question of 'middle latitude' does not arise, saying merely that latitudinal distances and course distances must be measured along the meridional line 'in those latitudes', and that he will call these 'proper distances' to distinguish them from longitudinal distances.

The following example indicates Gunter's method: 'By one latitude, distance and difference of longitudes, to find the difference of latitudes. Open the sector at right angles, so that the Line of Lines is on the inner edge. Set off the given difference of longitude along one line from the centre. Open a pair of compasses to the *proper* distance (i.e. measured along the meridian scale) and putting one foot on the term of the D. long. turn the other foot until it meets the other line of the sector. The line cut off gives the proper distance of the latitude required, and this set against the meridian scale gives the D. lat. in degrees.' In each example the error that would arise supposing the plain chart were used is stated, in this case a D. lat of $2\frac{1}{2}°$ instead of 5°. The same problem is worked out graphically with ruler and compass on the chart, while in his tractate on the cross-staff he gives the trigonometrical solutions. Here he uses mid-latitude, which he terms sufficient for seamen's use. For example: The longitude and latitude of two places being given, to find the rhumb:

> 'As the difference of latitude
> to the difference of longitude
> So the cosine of the middle latitude
> to the tangent of the rhumb required.'

Other trigonometrical formulae are given for data likely to be required by the sailor, e.g. the Sun's amplitude and azimuth when the latitude and declination are known. Gunter preferred to find the variation by comparing the observed and calculated azimuths of the Sun, and discovering it to be little more than 6° E. he went down to Limehouse, where Borough had found it $11\frac{1}{4}°$ or $11\frac{1}{2}°$, and took eight observations there at the summer solstice (1622). The difference from William Borough's observations (made there in 1581) was confirmed, and this led to doubts which about ten years later were confirmed by Gunter's successor Henry Gellibrand. For he himself died in 1626; not, however, before he had devised another admirable help for seamen. He put the logarithmic scale upon the yard-long rod of his cross-staff, calling it the Line of Numbers, and taught how to use it with the help of a pair of compasses to avoid tedious arithmetic. 'Gunter's Line', as it was called, was taken into use by sailors, who would not give it up even when the much more convenient slide-rule was invented by 1654. The example the author gives relates to the ship's way: say it was measured as 88 ft. in 15 seconds, then

> 'As the time given is to an hour:
> So the way made, to an hours way.'

Extend the compasses on the line of numbers from 15 to 3600, then the same extent will reach from 88 to 21,120. Hence the ship's way is 21,120 ft. or 4·224 miles of 5000 geometrical feet. This leads him on to question the acceptance of 60 miles of

16—H.A.

5000 ft. as the measure of the degree: 'Comparing several observations', he says, 'and their measures with our feet usual about London, I find that we may allow 352,000 feet to a degree.' From this he goes on to suggest a re-knotting of the log-line with the introduction of a decimal system of units. The matter of the degree and of the line was to be handled again a few years later, and in a more practical fashion, by Richard Norwood. Gunter's idea that all nautical computations should be carried out in terms of the degree of a great circle was a sound one, but not his attempt to substitute centesmes, or hundredth parts of the degree, for the familiar minutes. The relation of a mile to a minute of arc was too convenient a one to be sacrificed, and the nautical mile parted company with the geographical mile practically without discussion. It was a minute of arc, whatever length on the ground this might prove to be. That the Earth was not a perfect sphere was a fact not yet dreamt of.

Richard Norwood, who was about ten years younger than Gunter, had gone to sea as a youth and taught himself practical mathematics from the textbooks. It was during the period when he had settled down in London as a teacher and land-surveyor that he took the opportunity of business in York to measure with his ordinary equipment the distance between London and the northern city. This he compared with the accepted latitudes of the two cities, and obtained the result of 69 miles 14 poles to a degree, which was surprisingly accurate in the circumstances. He published his figures in 1637, in a textbook of navigation which he entitled: *The Seaman's Practice: containing a Fundamental Problem in Navigation, Experimentally Verified, viz: touching the Compass of the Earth and Sea, and the Quantity of a Degree in our English Measure, also to keep a Reckoning at Sea for all Sailing etc.* Here he proposed the re-knotting of the log-line at 50 ft. instead of the customary 42 (7 fathoms). Run with a half-minute glass, this meant that a knot was the equivalent of a sea-mile of 6000 ft. in an hour. He cast away some odd feet in his mile on the old principle that it was best for the sailor to believe himself nearer land than he actually was. Instrument-makers made the new line, but all through the century there were sailors who preferred the old one, meeting its error by altering the sand-glass so that it ran three seconds short of the half-minute. Indeed, it is almost incredible to learn that in 1727 the French naval captain Radouay was still complaining that sailors used a line knotted at 42 ft. instead of the $47\frac{1}{2}$ ft. that would agree with Picard's measure of the Earth made in 1672. The Minister Maurepas subsequently laid down 47 ft. 7 in. as the official distance of the knots, but there were more complaints in 1781 of the continued use of the old line. In Spain and Portugal (and among many French sailors) the log-line was not used at all, and the traditional rules for estimating the ship's way remained in force. The following table, found in a rutter for India of about 1604, is typical:

Com vento quanto em pôpa 43–45 [leagues to a *singradura* or day's sail]
,,   ,,   ,,   pela bolina 38–40
Com vento teso em pôpa 36–38
,,   ,,   ,, pela bolina 32–34
Com vento esperto em pôpa 33–34
,,   ,,   ,,   pela bolina 28–30
Com vento fresco em pôpa 30
,,   ,,   ,,   pela bolina 25
Com vento galherno em pôpa 24–26
,,   ,,   ,,   pela bolina 20–22
Com vento bonança em pôpa 18–20
,,   ,,   ,,   pela bolina 16–17$\frac{1}{2}$
Com vento calma em pôpa 14–16
,,   ,,   ,, pela bolina 12–14
Com vento quanto a nau governes em pôpa 10
,,   ,,   ,,   ,, ,,   ,,   pela bolina 8

Ter-se-ha respeito ao mar: se é châo ou picado, que detenho a nau e ao ir velejada ou não e tambem espedir a nau por bordo e ao ir mui carreganda ou leve.

In brief, winds of seven strengths, from a high wind to mere steering way, were recognized, which might be on the poop or on the quarter, giving the ship way accordingly. But regard must be had also to the sea and to the canvas carried by the ship. That was all.

The Spaniards and Portuguese continued also to teach the use of the plain chart, although Manuel Pimentel's *Arte de Navegar* of 1699, which continued in use far into the eighteenth century, had a table of meridional parts and some brief account of the 'reduced chart' (i.e. Mercator) placed in an Appendix. This author considered unnecessary such refinements as correcting for horizon dip and for refraction (tables for which he said were to be found in Dutch manuals), since they cancelled one another out, and the old-fashioned instruments which continued in use among Iberian sailors received no critical consideration at his hands.

The discovery of the secular variation of the variation of the magnetic needle appears to have made little impression on the maritime world. Indeed, Professor Gellibrand, when he announced its confirmation in 1635, merely detailed his observations without comment. The yearly alteration was a slow one, and sailors did not think in terms of minutes of arc, nor indeed did men of learning, apparently, grasp the fact that an observation of the variation ought consequently always to be dated. Fournier, in his great *Hydrographie* of 1643, quotes extensively from the writings of William Borough, Edmund Gunter and Henry Gellibrand, and reports that at Nuremberg, where the great instrument-maker Georg Hartmann had made the variation 16° E. about a century earlier, he learned that it had now fallen to 7°, or about five minutes' diminution in a year, but he gives no strictly reliable data. Father Riccioli in his *Hydrographicus* of 1661 listed over four hundred observations but gave dates for only about half a dozen. Gellibrand's results were, however, seized upon by a young Englishman who taught navigation down at Chatham Docks and he was to make a lifetime study of them, drawing up a scheme by which (so he believed) variation could be used to determine the longitude. This man was Henry Bond, who took on the editing of the *Seaman's Kalendar* after John Tapp died. When he was an old man his scheme for the longitude was examined by a Royal Commission, but this event belongs to the later period when the English 'Royal Society' concerned itself with navigation problems.

Bond, as himself a teacher, closely followed all that other teachers wrote, and in the *Kalendar* he commented on the reformed degree, writing:

'Here note that although the Author [Tapp] hath set down 20 English leagues yet it must be 20 such leagues as answer to one degree of the meridian; and therefore the knots on the log-line must be 50 ft. asunder at least, according to the experiment made by Master Richard Norwood, which experiment was formerly verified by practice at Sea by Captain Thomas James in his voyage to the Northwest, as we may see in the 7 page of his Journal. . . . But because many will hardly be drawn to alter their knots from their old form, therefore if any one will multiply 112 by the knots run out in half a minute, the product cutting off two figures on the right hand shall be the number of leagues run in a watch according to Master Norwood's experiment.'

Norwood gave a very clear exposition of great-circle sailing, the method for which in effect was to replot short lengths of the circle on a Mercator chart, and sail on a succession of straight rhumbs. In his diagram (Fig. 28) AB is the great circle between two places, so that the angles PAB and PBA are found. The triangle is then divided by meridians PX, PY, PZ, etc., at a chosen angle from PA, say 10°, when the triangles PAX, PXY, PYZ, etc., can be solved. This gives the bearings and distances of the segments AX, XY, YZ, etc., of the great circle, whose latitude and longitude are known. P is the Pole, so that the arcs PA and

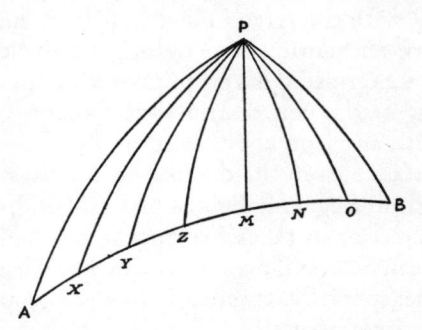

Fig. 28. Great-circle sailing by Richard Norwood's rule.

PB, and the angle APB (D. long.) are known. This triangle can be solved. These short sections may be replaced without noticeable error by straight-lined rhumbs. But sailors dismissed great-circle sailing as too difficult.

Fournier's book gives a good picture of the state of navigation in England, Holland and France in his day, for the three countries watched one another's methods closely, and the English used Dutch charts and 'waggoners'. In the stern of a great ship there were four decks above the bilge, the bottom one used for stores, the second for the guns, the third for the great cabin and the binnacle, the fourth for the master's cabin and the pilot's station. This fourth deck was the *dunette* or poop, and from it there was a direct drop or companion-way into the captain's cabin. The tiller from the ship's rudder passed across the gun deck, under the floor of the great cabin, and the whip-staff by which it was moved came up behind the binnacle, or bitacle as it was originally termed.

'En ce mesme [says Fournier of the third deck] est l'Habitacle, appelé des Marseillois la Custode ou Girole, où sont trois niches des Armoires, et parfois quatres: en l'une est la Lumière, en l'autre la Boussole, Compas ou Quadrans de Mer, en la troisième l'Òrloge ou Poudrier. S'il y en a quarte, on y met deux compas. On met pour ordinaire dans un Vaisseau huict Compass, vingt-huit Horloges d'Escales d'œufs, une horloge de trois ou quatre heures, et deux d'une minute chacune.'

This use of egg-shells in the hour-glass was probably not universal, but it was known to Richard Polter. Fournier had, however, sailed with the Grand Fleet of France, and his information is therefore authentic. In detailing the duties of the pilot (which included exercising extreme care over the soundness and correct working of the compass) he notes that he will not go to the binnacle with any iron about him, and will see that the guns are kept at a distance from the compasses. Deviation was, in fact, well understood. In England, France and Holland, Fournier adds that every skilled seaman takes his turn at the helm, whereas in the Mediterranean Sea (and presumably in Spanish and Portuguese ships), only the specially trained helmsmen could do so. He explains the log-line, as used by the English, but this method of finding the ship's way was evidently not in general use among French pilots, although their skill in 'estime' or dead reckoning was their chief qualification. Both the French mathematician Herigon and the Dutch Snellius had earlier praised the instrument, although, as Fournier pointed out, it was essential to note wind and current movements, since it had to be assumed that the log-board remained stationary as the line ran out.

The daily Journal had now long been stereotyped, and as Fournier set it out had eleven columns, under the headings: (1) Day and month, (2) Rhumb of course, (3) Wind, (4) Quality of wind, (5) Glasses (hours), (6) Distance (leagues), (7) Estimated latitude, (8) Difference longitude, (9) Observed latitude, (10) Compass variation, (11) Accidents. In John Davis's Journal, or Traverse-Book as he called it, made for his voyage of 1587, which was printed by Hakluyt, there were only seven columns, corresponding to numbers 1, 2, 3, 5, 6, 9 and 11 of the fuller Journal. The 'true course, distance and latitude' as worked from the successive traverses, i.e. columns 7 and 8, were included, together with occasional observations of compass variation, in the last column of miscellaneous observations, here termed the Discourse.

The Journal of Wm. Haddock, commissioned to the *Cornwall* in 1696, has only eight columns of which six correspond to those in the French list. They are: (1) Day and month, (2) Wind, (3) Course, (4) Distance in miles, (5) Latitude corrected, (6) Longitude corrected, (7) Bearing of headland last seen, (8) Remarkable observations and accidents. Hence the traverses must have been worked out and the ship's position in latitude and longitude determined before entering up the Journal. Similarly the distance covered was worked out from the readings on the log-table, and the hours (or 'glasses') on each rhumb, therefore, are not entered. The log-table ran: Hour, knots, feet, course, wind, leeway (in points). Lieutenant Edward Harrison, who had entered the navy from the East India service for the French wars, wrote in the same year 1696 that a printed form for the Journal was now handed out by the Admiralty, and that a column ought to have been added for compass variation. But he had served on ships of every rate, and had never set eyes on an azimuth compass in one of them (Plate XX).

In France, when Fournier wrote, there were officially appointed pilot-hydrographers at the various sea-ports who were obliged by a regulation of the Minister of Marine to give instruction three times a week to anyone seeking it. In England all instruction was in private hands, apart from the odd post at Chatham Docks, which was not necessarily filled, nor with a competent man. Fournier mentions the provision of teachers in connection with what was an outstanding controversy of the seventeenth century—the gentleman-captain as opposed to the 'tarpaulin', or captain who had been bred to the sea. He was of the opinion that the former was the better leader aboard a naval vessel, but that he should learn something of navigation. The same view was held in England, although some famous captains did, in fact, start their careers as ships' boys, for example Sir Cloudesley Shovell.

2. THE LEARNED SOCIETIES

Towards the middle of the seventeenth century, a new factor was to influence navigation methods. Groups of savants both in France and in England began to meet together in private to discuss the new ideas and new discoveries in science which were so at variance with ancient thought, and these Invisible Colleges, as their members termed them, were later transformed into learned Societies. The French *Académie des Sciences* (1666) was a body of scholars paid by the king, who were set definite problems to solve, and work to be carried out, according to programmes drawn up by the king's Minister. The English *Royal Society* (1662), although enjoying the king's patronage, was a body of members elected among themselves whose programme was largely unplanned, depending upon the interests or enthusiasms of individual members or of the officers. In fact, it might be said that the one society was professional, the other amateur, and each mode of organization had its advantages, although in respect of the first great task carried out by the *Académie* it looked as if professionalism was the more efficient. This task was one fundamental to cosmography and to the chart, namely the precise determination of the length of the degree of the meridian. The Royal Society repeatedly told their Curator, Robert Hooke, to measure the degree, and at a later date they offered Edmund Halley £50 or fifty copies of a valuable book if he would carry out the task. But it was one which required time, man-power and equipment, that is to say it lay quite beyond the scope of an individual. Under Colbert's direction it was undertaken by the academician Picard, with his assistants, and took him nearly two years. This was during 1669–70. Before this date, it should be noted, the conditions for making accurate observations had greatly improved. Both scale-makers and optical-instrument makers had gained in skill and precision, and several new inventions had been made. The military engineer Vernier, for example, had invented the device which bears his name for subdividing the small divisions of a scale. The vernier is a little plate which runs smoothly along the edge of the scale, carrying a second set of divisions each of which (where a decimal notation is being used) is 0·9 of the smallest division 0·1 on the main scale (that is to say each vernier division is 0·09). The edge of the object being measured is aligned to the left-hand edge of the vernier scale. If this does not also coincide with a division on the main scale, the eye is carried along the latter until a vernier and main-scale division are found which coincide. Suppose this is 4 on both scales, then the reading is 0·04 more than the nearest main-scale reading to the left of the object being measured. The reason is that $4 \times 0·09 = 0·36$, while $4 \times 0·1 = 0·4$, so that the vernier must have been moved $0·4 - 0·36$ from the reading on the main scale nearest to the object, i.e. 4 hundredth part or 0·04. A second invention was the micrometer, first devised by a young Englishman, George Gascoyne, but later improved by Picard and the *Académie* secretary Auzout. The micrometer consists of a screw through the side of a telescope tube which parts or closes a pair of points (in Picard's case two very fine silk threads) in the focus of the object glass. The screw (which has a very fine thread) turns also a pointer on a dial, and so the distance between two objects seen through the telescope can be measured. Even without the micrometer the telescope was improved by having cross-wires or cross-threads placed in the focus of the object-glass, so that the object observed could always be brought to the same spot. Telescopic sights, besides, were added to surveying instruments, and greatly improved them, despite the obstinacy of the great Hevelins who insisted that the bare-eye observations were always the best.

The measure of the degree effected by Picard was carried out by triangulation, the method followed by the Dutchman Snellius in 1617 when he measured the distance and the bearing between Alkmaar and Bergen op Zoom. By solving the triangle he found the meridian distance between the two towns, whose latitudes he also measured by the Pole Star. But Picard first measured an accurate base-line of 5663 fathoms by means of iron rods, and then continued northwards through a system of seventeen triangles, making a second direct ground measurement between the thirteenth and fourteenth of them as a check. The whole series stretched from Malvoisin to Amiens, and at three points the latitude was determined. For these astronomical observations an instrument of 10-ft. radius was set up, while for taking horizontal angles the arc of the instrument had a radius of $3\frac{1}{6}$ ft., both having diagonal scales. Even so, the observers could not avoid an error which they put at 2 seconds, and as this represented 22 fathoms on the ground, the result was held to be unsatisfactory. The figure found was a degree of 366,669 English feet, or 69 miles 783 yd., which differed by only 706 yd. from Norwood's very rough-and-ready measure between London and York. This coincidence was perhaps fortunate as influencing English assent to the better figure even outside scientific circles, and making teachers of navigation more emphatic about the 'false' log-line. Meanwhile the Frenchmen continued their meridian measurement to the south of France, and from further surveys redrew the map of the country in such a way that striking longitude errors were revealed. The king was startled at the apparent reduction made of his kingdom from east to west, and not wholly pleased.

The longitude question was of course in every man's mind, but it will be considered separately in the next chapter, for there were other matters relative to navigation which called loudly for consideration, even leaving the chart and longitude aside. One of the factors making dead reckoning so uncertain still, however, remained baffling, namely the allowance necessary for leeway. Some sailors trailed a lightly weighted cord from the lee-side of the poop and watched its motion. Others used a rule of thumb of the type employed by Spanish sailors for estimating the ship's way, the English version of which an old sailor explained to William Jones, a teacher of navigation in the reign of Queen Anne. If the wind blows east-north-east and the ship's stern is north allow one point. If the wind blows so hard that one top-sail must be taken in, then three points, and so forth, ringing the changes on wind and canvas. But for the other great impediment, 'unknown currents', the outlook was more hopeful. It was already ordinary practice to put out in the ship's boat, let down a sea-anchor, and then throw out the log to find a suspected local current, but it had also been pointed out that in the main currents followed the winds, and hence Edmund Halley's attempt at a systematic survey of the steady wind systems (1686) promised improved knowledge. It was furthermore of importance in itself. Halley had had ample time to study wind behaviour when as a young man he spent a lonely year on St. Helena, mapping the southern stars.

And there was abundant descriptive material besides in the literature of sailing. He was able to define with considerable accuracy the character of the trade-wind belts in the Atlantic Ocean, their limits and their shift with the seasons, as also the anomaly of the south-west winds which always blew in the Gulf of Guinea. It was these, he pointed out, which negatived the old view that the air all flowed eastwards under the influence of the *primum mobile*, the supposed turning heavens. They lent support, too, to his own thermal hypothesis, correct so far as it went, that air flowed in towards the hottest areas. In the Indian Ocean he described the contrast between the steady trades to the south of the equator and the monsoon system to the north of it, and he gives some account of the variable and often stormy conditions to be expected during the change of the monsoon. The different conditions between north and south he tentatively attributed to the distribution of the land masses, a heated area being developed in south and south-east Asia. The vast Pacific Ocean was as yet but little known. The monsoon area extended through the China Seas, but it was reasonable to suppose that in general over the open ocean the system would resemble that in the Atlantic. This was supported by the fact of the known ease with which the annual Spanish galleon went to and from the Philippines, keeping south in the trade-wind belt on the outward journey, and seeking the variable winds in higher latitudes for the return, just as was done in the Atlantic. However, in the map which accompanied Halley's historic paper (read to the Royal Society) the greater part of the Pacific Ocean was omitted since its meteorology was so largely speculative. This pioneer wind map was later used to illustrate William Dampier's useful *Discourse of the Winds* (1699), and appeared subsequently on many maps and globes with little alteration.

There was no material as yet for any general map of ocean currents, such as Samuel Pepys thought would be so useful, but Halley pointed out that if the strong ebb and flow of tidal streams in the English Channel was properly charted it would immensely improve the chances of English ships getting down the Channel without delay. For the stream might be used as a set-off against contrary winds. The result of this representation was that Halley was given a temporary Captain's commission in the navy (which did not please the lieutenant aboard) and during the closing years of the century made observations at sea upon which a new chart was based. This was published by Mount and Page, then opening their career as the great eighteenth-century chart-makers and publishers. It was entitled: *A Chart of the Channel between England and France with the flowing of the Tydes and the setting of the Currents, by Captain Edmund Halley*. An explanatory text, containing notes also on the variation of the variation (for there were magnetic variations on the chart) and warnings about the general errors of charts, was issued as *An Advertisement* in the public press, and was sometimes pasted on the chart by the publishers.

In the winter of 1699–1700 Halley made a longer oceanic voyage, particularly to observe the variation, at the command of William III, and as a result there was published a world map on Mercator's projection which was marked with isogonic lines, that is to say lines drawn through the points of equal variation (Plate XXV). This was the first isometric map (apart from some attempts to show depth contours) to come into general use, and its appearance was therefore a signal cartographic event, quite apart from its intrinsic value to sailors. The map was reprinted again and again, in French and Dutch as well as in English editions, and after an interval the data were revised in 1744 and again in 1756, this last edition being prepared by two writers on navigation, William Mountaine and James Dobson, who also supplied an explanatory pamphlet on their methods. This appeared in the *Proceedings of the Royal Society* in 1757, for both were Fellows, and in it they stated that for the earlier revision Messrs. Mount and Page had employed Charles Leadbetter, a well-known teacher and writer, who had used 1100 new observations made by one Robert Douglas, a teacher of navigation aboard his majesty's ships. They themselves, however, had prepared a table of 50,000 figures which were adapted to every 5° of latitude and longitude over the most frequented oceans. Moreover they had the advantage of a recent (1750) great improvement in magnetic compasses. These instruments had been neglected 'until of late years the judicious Dr. Gowin Knight, F.R.S. examined into their fabric and construction, employ'd his Magnetic Knowledge towards their Improvement, and has now reduced them to a considerable degree of Perfection, as Experience has sufficiently evinced, more especially since they have been approved by the Commissioners of the Navy, and ordered into Use in all His Majesty's ships of War. These Compasses are not only fitted for Steering, but also for taking the Sun's Amplitude and Azimuth by adding an easy and simple Apparatus for those purposes: and are made by George Adams, Mathematical Instrument Maker to his Royal Highness the Prince of Wales, and before they pass out of his Hands, are examined and attested by the said Dr. Knight, whose Certificate is fixed to the Cover of the Box: without which they are not to be depended upon.' This new care for instrumental standards is worth remark. Dr. Knight used magnetized steel bars for 'touching' the needle instead of a lodestone.

A French edition of the revised map, dedicated to the Minister of Marine, was published by the engineer Bellin, in 1765, and he advised the reader that he had used the English tables, and that as the variation increased westward by nine or ten minutes a year, the figures should be modified accordingly. This rule was not, however, without exception. In Madagascar, for example, the reading had apparently not altered in a hundred years. The isogonic lines drawn by Halley covered only the Atlantic Ocean south of Lat. 50° and the Indian Ocean, to which areas his voyage was limited. He had previously made a study of the general subject of the Earth's magnetism, and in the second of two papers had propounded the view that in addition to the two main magnetic poles there were two others lying obliquely to them which slowly shifted. Bellin's map showed, besides the magnetic variation, the steady and variable winds (after Halley), but the convention employed, little cherub's heads puffing out the wind disposed here and there over the map—and in the case of variable winds four of them blowing in one another's faces—was much inferior to that originally used by Halley. 'That the Reader', he had written, 'may form the better Notion, we shall add a Figure [map] presenting to the Eye all the Quarters and Points of all the Winds. The Limits of each Tract are marked with pricked Lines, as well in the Atlantic, whether they separate the variable Winds from the constant, as in the Indian Ocean, where they also' separate the different Monsoons from one another. The easiest way of marking the Quarters of the Winds seemed to be by a series of little sharp-headed Lines, pointing alternately to the Parts of the Horizon from whence the Winds blow.' Actually the engraver did not make the short lines sufficiently 'point-headed' to show whether the wind was blowing, say, north-east or south-west, but the series of short lines give the sense of a flow of a stream of air, of converging streams, and alternating ones, and assuming a general familiarity with trades and monsoons, the picture is a vivid one. Later engravers added a few normal wind arrows but the development of map conventions was still in its infancy. Halley's Channel Chart, for instance, is drawn in plain-chart style with the conventional pattern of wind-roses and rhumbs, on a scale of five English and French leagues to an inch, and has the explanatory note: 'In this Channel Draught the smaller figures are the Depth in Fathoms. The Litteral or Roman Figures shew ye Hour of High Water, or rather ye End of the Stream that sets to ye Eastward on ye Day of ye New and Full Moon. Add therefore ye time of the Moon's Southing or Northing to ye Number found near ye place where your Ship is, and ye Sum shall show you how long ye Tide will run to ye Eastward. But if it be more than 12 subtract 12 therefrom. The Direction of ye Darts shew upon what Point of ye Compass ye strength of ye Tide sets.' The magnetic variation, $7\frac{1}{2}°$ W., is written in mid-Channel in three places, presumably where it was observed, and on the world chart the isogonic lines here run generally east–west right across to America. Besides depths (by convention those at half-ebb), there are indications of bottom, especially near the Channel entrance, but these and the centuries-old symbols of stippled sand-banks, crosses marking rocks, and anchors at anchorage were too familiar to need explanation. The publishers add this appeal: 'All Masters of Ships and others, who shall have opportunity to observe ye Depths, with Certainty in respect of ye Place, are desired to communicate them to ye Publisher hereof.' There was as yet no official source of information for chart-makers, except from the Trinity House surveys of the Thames, although recently a one-man survey of the English coasts had been made and published by Captain Greenvile Collins with some scanty help in money and ships from the Crown.

Gradually, of course, the general scientific background of nautical problems was improving. The French Academician, de la Hire, for example, had produced a table of atmospheric refraction which did away with the idea that this differed according as Sun or stars were being observed, while the English instrument-maker and experimenter, Francis Hauksbee, senior, demonstrated that there were variations in the figures according to atmospheric conditions. Towards the end of the seventeenth century, again, the significance of the rise and fall of the barometer in relation to weather and to weather forecasting began to be understood, and Robert Hooke devised a marine barometer for use at sea, although perhaps he was optimistic in suggesting that: 'The observations of the weather may perhaps in great measure be timely enough discovered by the inquisitive and diligent mariner.' Nor should we overlook Isaac Newton's quiet announcement of the composition of white light, although it was not until 1758 that the optical instrument-maker John Dollond invented the achromatic telescope.

Above all, the use of the telescope and the introduction of the pendulum clock had made more accurate astronomical tables possible, and while Tapp's *Kalendar* and the better private annual almanacs (such as John Parker's) were still relied upon in England, members of the French Academy produced under Royal patent the greatly superior *Connoissance des Temps*. This yearly publication, instituted in 1678/9, was taken over by the *Bureau des Longitudes* in 1798 and has appeared continuously down to the present day, so that it must rank as the first Nautical Almanac. The tables are for the meridian of Paris, but a list of longitudes of nearly ninety other towns is given in terms of the difference of time in minutes of time from Paris (e.g. London 9′ W.) with instructions on how to adapt the tables to other places of observation. Of particular interest to sailors are the full tables of the Moon's motions, with a long list of Establishments in France and neighbouring lands, and a detailed explanation of how to find the hour of full sea on any particular day in the calendar. For telling the time at night data are provided of thirty night stars which do not set in France, so that their meridian passage could be compared with the Sun's right ascension taken from the appropriate table. The equation of time and the method of setting and regulating a pendulum clock are explained in detail, but such a clock could not be used at sea, as experiments early showed. Huyghens had believed that his marine pendulum clock would solve the longitude, but longitude still continued to defy solution. Ships continued to be wrecked in consequence, and sailors to die.

XI

The Longitude Solved

1.

As far back as the thirteenth century Peter Peregrinus, the great experimenter of his age, had been laboriously shaping, spinning and testing his lodestones in the belief that an instrument could be designed which would turn about precisely with the heavens each day through the sympathy of all its parts. All that was necessary was to get the magnetic stone exactly cut and poised. Such an instrument would be worth a king's treasury, said his friend, Roger Bacon, for it would solve all astronomical problems, and provide an infallible timepiece. And not many years later, one Robert the Englishman, who was lecturing in Paris on Sacrobosco's *Treatise of the Sphere*, remarked that there were people who believed that they could make a mechanical clock which would keep perfect time, and so supersede the astrolabe (by which time was found), although he himself considered it a vain notion. And so, indeed, it remained for the next five hundred years. Nevertheless by the early sixteenth century the making of clocks had so far improved that Gemma Frisius of Louvain, who was in the service of the king of Spain, thought it worth while to point out that a trustworthy clock, if carried to sea, would solve the difficulty of finding the east–westing on the new oceanic voyages. This was said in 1522, and when, some thirty years later, the English were full of ideas for new exploration and discovery, their historiographer Richard Eden (as has already been mentioned) translated the passage from Gemma's book, and printed it in his *Decades of the Newe Worlde*, 1555. Henceforward the idea was present in men's minds; indeed a friend of Eden's, William Borough, is supposed to have tested it. But the watches of the time gained and lost up to fifteen minutes a day, and needed frequent winding, whereas what is required is a timepiece which will be accurate to seconds over a period of weeks. For our globe whirls around at such a rate that time alters in the proportion of almost four minutes for every degree of longitude, so that a timepiece losing twelve minutes in time would falsify the longitude by 3°. Contemporary charts, of course, contained very gross errors, as a quite casual examination at once reveals, errors up to 7° or more; yet even these could not have been corrected by the method, more than once proposed, of carrying a sand-glass which emptied itself in twenty-four hours, and must then be promptly turned, or by carrying water-clocks. Quite apart from the effects of the ship's motion, those of temperature change were over-looked. Even when, by the eighteenth century, clock-makers were able to produce timepieces of very considerable accuracy, they found themselves baffled by the fact that owing to the expansion and contraction of metals with temperature change, not to speak of the behaviour of the lubricants they used, there was still a degree of unreliability in a spring watch or clock that they could not overcome. It is little wonder that many people turned again to consider the heavenly bodies as timekeepers, and particularly to the swift-moving Moon.

The cry of the sailors was heard continually. To find the longitude they had to catch at straws. Here, for example, was Abraham Kendall, the man whom his fellow-mariners distrusted, but who was said to have taught young Robert Dudley, Leicester's son, enough navigation for an admiral. Crossing the Atlantic in 1594, he put into practice the widely accepted theory whereby the intersection of the line of magnetic variation with the latitude supposedly gave the longitude. He enters in his Journal: '9° 56' N, Long. 337° E. [of the Azores]. We are 640 leagues of a great circle from the Cape [Blanco]. Here we began to see some birds of the Indies called *forcedos* [frigate birds]', and presently: 'Lat. 9° 30', Long. 335°. We saw as a sign we were nearing America some great birds like crows, but white with long tails.' Birds began to settle at night on the rigging, while the sea water was considerably whiter. 'These are signs of the neighbourhood of the coast.' And it was just the same on the homeward run: 'Bermuda was distant 110 leagues of a great circle, and we saw frigate-birds and sea-mews. The current ran to north-east [the Gulf Stream] and carried with it weeds from the rocks of the Indies. And when one no longer sees such weeds it is a sign that one passes C. Race, Long. 344° 10' from Pico of the Azores.' He was quite right, of course, to watch and interpret the birds, and the weeds, and the colour of the water, for that the magnetic variation could indicate longitude was only a dream, though one which recurred again and again. Perhaps the last time was when Dr. Johnson helped the aged Zachariah Williams to publish a pamphlet on the subject in 1755, when the chronometer was actually in being. So distinguished a quartette as Dr. Gilbert, Thomas Blundeville, Edward Wright and Henry Briggs were responsible in 1602 for a scheme to use the dip of the needle, discovered by Robert Norman, as a means to find the latitude when the sky was overcast, but they did not go so far as to suggest that it would be useful for the longitude. This was left for Henry Bond, and his triumphantly titled book, *The Longitude Found*.

Bond's story is a curious one. As already related he set to work to study the variation of the needle when Gellibrand was first able to announce its secular change, and in 1638 he foretold in Tapp's *Kalendar* (which he was then editing) that the variation (then east) would be zero in London in 1657, and would subsequently slowly increase to the west. This indeed did take place and made people inclined to listen to him when he drew up a theoretical table of the magnetic elements to be used with a dip needle to find the longitude. The king set up a commission of six (which included Robert Hooke) to examine the method, and they decided to approve it (although Hooke knew it to be worthless) not for its merits, but rather to spike the guns of a French pretender, the Sieur de St. Pierre. The same committee had been asked to assist this gentleman and report on the success of his method, and he demanded various astronomical data from them in a decidedly arrogant fashion. What he proposed was in fact what Jean Rotz claimed to have done, namely to measure the Moon's appulses to the fixed stars. Certainly the conditions were less unfavourable than they had been more than a century earlier, but it so happened that young John Flamsteed, who had been working on the Moon's motions in Derbyshire, was on a visit to London, and he was co-opted on to the committee. It was Flamsteed who drew up the report. The Sieur's method (he wrote) 'was entangled with uncertain refractions and dubious parallaxes', while the current lunar tables had errors up to eight or nine minutes, or even (according to his own observations) up to a quarter of a degree. This was the report that led King Charles II to found the Royal Observatory at Greenwich and put Flamsteed in charge of it as Astronomer Royal. There the young clergyman grew old, spending a whole lifetime on drawing up a new catalogue (which did not appear until he was dead) of the fixed stars, but publishing from time to time new tide-tables based on his observations of the Moon.

The French Royal Observatory had been completed near Paris in 1672, three years before the English one at Greenwich, and there the Italian astronomer, J. D. Cassini, drew up tables of the occultation of Jupiter's satellites. For since the Academicians had pensions from the king, eminent foreign savants could be invited to join their company and work for France, as Huygens did, and Olaus Römer. That Jupiter had satellites was discovered as soon as the telescope was turned on to him, and Galileo at once realized that their eclipses, taking place almost nightly, would serve far better than the eclipses of the Moon for the determination of longitude. Once a precise table, timing each eclipse, was drawn up for a standard meridian, then the difference between this and the local time of the eclipse would give the longitude. But would this serve for longitude at sea? Could a sailor use a big telescope on board ship? And could he time an occurrence to seconds or even to minutes? However, these objections had not the same force on land, and not only Cassini, but Robert Hooke busied himself with making precise tables of the occultations. And Römer, in the course of his observations for the same purpose, discovered that light had a finite velocity, and did not arrive instantaneously as was generally thought. When the Earth was in the part of its orbit most distant from Jupiter the satellites reappeared a mere trifle late. A table for the first satellite was published in the *Connoissance des Temps* from 1690 onwards, and it might have been thought that the invention of the Newtonian (or Gregorian) reflecting telescope about 1670 would allow of observations at sea. For this telescope is relatively short and so could be held steady as a ten- or twelve-foot instrument could not. But the polished speculum or reflector soon clouds and tarnishes with damp, thus becoming useless. Suggestions were made from time to time—and even tried out—that the observer at sea should sit in a heavy swinging chair, which would remain steady as the ship rolled. Jacques Besson, the royal instrument-maker (and Richard Eden's friend), had had that notion and pictured it as far back as 1567, but it did not work; such a chair moved irregularly. Jupiter's satellites proved a disappointment, and of course this planet is not always visible in the night sky.

The invention of the perfect timepiece likewise remained a will-o'-the-wisp. Robert Hooke maintained that he had made a spring-watch, when he was at Oxford just before the Restoration, of a sort that would solve the longitude, but when his friends tried to help him to patent it he took offence for some obscure reason, so that it was never given a trial. But that he made improvements in watches is generally agreed. A few years earlier, in 1657, Christian Huygens had invented the pendulum clock, a clock that is to say which, whether driven by weight or spring, was regulated by the isochronous swing of a pendulum. Two years later he invented a marine form of his clock, in this case spring-driven, with a half-minute pendulum. It was heavily weighted and swung in gimbals. His experimental work had been done in the Netherlands, where there were many English exiles during the Interregnum, so that it came about that the clock was tested on English ships. The first reports on the pair of clocks taken to sea in 1664 by Major Robert Holmes were highly favourable, but the master of the ship told Mr. Pepys privately that the two clocks differed from one another sometimes in one direction, sometimes in the other, and Hooke, too, threw doubt on the matter. Huygens himself, however, was still confident, and the French king, Louis XIV, had two of his clocks, in this case weight-driven and regulated by cycloidal pendulums, sent out with an expedition to Candia. The result was claimed to be good, and in 1669 the inventor prepared a pamphlet of instructions on the care and use of such clocks at sea. But in the event they proved disappointing. The pendulum could not ride a storm. And furthermore the disconcerting discovery was made by Richer in 1672 that the seconds pendulum did not swing true when carried to the equator. It had been supposed that the length of a pendulum which swung to and fro every second would serve as an absolute unit of length, easily found and changeless. But in actual fact the period of swing depends not only on the length of the suspending cord, but on the gravity constant g, which in its turn varies from pole to equator. The pendulum fell, therefore, under suspicion as a portable timekeeper, although pendulum clocks were extremely valuable in themselves, and introduced the public to the fact (very shocking to some) that the Sun (and therefore the sun-dial) does not keep precise, or rather uniform, time throughout the year. The 'equation of time' began to appear in every Almanac, whereas it had hitherto been of interest only to astronomers, and sun-dials were gradually relegated to the status of garden and house ornaments.

During the latter part of the seventeenth century most astronomers had come to accept the position that the longitude could only be solved by means of precise observations of the Moon: and the necessary degree of precision was set by the fact that an error of 5′ in reading the Moon meant an error of $2\frac{1}{2}°$ in the longitude. And such an error might arise from the imperfection of the tables for the standard meridian with which the local observation was compared; from the imperfection of the instrument used to take the observation; from the imperfection of the figures used to correct observations of Moon and star for refraction; from the imperfections of the figure used to correct for the Moon's horizontal parallax; and from combinations of all these. At sea it was necessary to add also the motion of the ship, and the lack of training of the observer, which would show itself in the carelessness and imprecision of his observations, and his incapacity for the necessary mathematical calculations. It was difficult, with all this long list of impediments, to know where to begin.

As regards the last difficulty listed—the need for a mathematical sailor—it remains to this day. The Spanish system instituted at the Casa de Contratación in 1502 gradually perpetuated fossilized and old-fashioned methods which were followed also by the Portuguese. The French system of State-paid 'professors' in all the ports gave more opportunity for variety and initiative, but sailors were not forced to seek instruction. In England everything was left to private enterprise, and anyone who chose could set up as teacher. The first step in advance was made when Charles II instituted a mathematical school for forty boys at Christ's

Hospital. But here there were difficulties at once, since a master was hardly to be found who could satisfy the practising sailors on the one hand and the mathematicians on the other. And, furthermore, the Governors of the hospital asked what was the need for a lot of new-fangled teaching of a sort that heroes like Drake and Hawkins had done very well without? Sound advice, however, came from Isaac Newton, consulted in 1694 about an old and a new syllabus. The first, he wrote, was no more than could be learned parrotwise, 'whereas the Mathematicall children, being the flower of the Hospitall, are capable of much better learning, and when well instructed and bound out to skilful Masters may in time furnish the Nation with a more skilful sort of Sailors, builders of Ships, Architects, Engineers, and Mathematicall Artists of all sorts, both by Sea and Land, than France can at present boast of.' And Mr. Pepys, as Secretary to the Admiralty, had helped matters forward by instituting the naval lieutenant's examination (which included navigation) in 1677, while by 1702 there were naval schoolmasters aboard ship, examined and licensed by Trinity House. By about the same date, too, a better type of master had been found for the mathematical school, while other similar schools were being founded, e.g. at Rochester. But although this meant that a basic mathematical skill was becoming the usual thing in the navy, it did not mean that a ship's officer, whether naval or marine, could make delicate instrumental observations of the Moon.

Instruments (apart from the costly giants housed in observatories) were, indeed, as much in default as competent observers. Forty years before the importunate Sieur de St. Pierre had given the spur to the establishment of Greenwich Observatory, another Frenchman, one Morin, had confidently believed that he had solved the longitude by the Moon. His spherical trigonometry was sound enough, and he was unaware of the defects of current tables, while as for an instrument, he proposed to use a quadrant of 2-ft. radius. A vernier affixed to the scale would read quarter degrees, a telescopic sight was to be attached to the alidade, and the instrument was to be weighted with 100 lb. of metal to keep it steady at sea. Morin even showed competently how to reduce the apparent positions of the heavenly bodies to their true positions by corrections for refraction and parallax—in theory at least. But the French judges refused him any reward, and from Holland, too, he gained no prize as he had hoped, although later Cardinal Mazarin pensioned him. Theory, in fact, was not enough.

The first step towards an efficient instrument was made by Robert Hooke in 1666. It was generally considered that the most practicable among the many possible observations of the Moon was to measure her distance from a fixed star, although Halley preferred to measure the moment of occultation of a star by her dark limb. It was to measure a lunar distance (as it was termed) that Hooke proposed to bring the one heavenly body and the other together at the point of a pin by observing one directly and the other by reflection in a mirror turned about on the alidade. The instrument was shown to the Royal Society in 1670, but Hooke only rarely followed up his ideas to a final practical conclusion and did nothing more. Isaac Newton, however, made an instrument (Plate XXI) with a fixed and a movable mirror attached to a quadrant, the former at the end of the telescopic sight, with the axis of which it made an angle of 45°, and the latter at the same angle to the fiducial line of the alidade. He brought an improved version of his instrument to the Royal Society in 1699, which gave Halley occasion to remark that he had used the earlier one for the longitude with success, but unfortunately both the instrument and a paper which accompanied it were put aside and did not become generally known until well into the following century. The reason was, no doubt, that the precise lunar tables and star lists awaited from the Royal Observatory had not yet appeared. There had, it is true, been some new lunar tables published in London in 1662. Their author was a mathematical teacher and custom-house clerk, Thomas Streete, and many people thought and spoke highly of them, including Halley who had admired Streete when he himself was only a youth. Indeed, in spite of Flamsteed's strictures on the tables, he re-edited and republished them in 1710, with an account of some observations of his own, made with a six-foot instrument such as he was sure that seamen could successfully handle at sea. Others did not however concur.

But within a very few years, as it happened, the situation was drastically changed. In 1707 the English fleet had met disaster on the Scilly Isles, and among the many drowned had been the

XXII. Hadley's octant (later sextant) was most generally used for bringing the reflection of the Moon down to the horizon, or for finding a lunar distance from a star of known position.

XXIII. Harrison's No. 1 Chronometer.

Mulgrave

July. 1775
Sat: 22.
When after two hours Calm, in the Latitude of 39..35 N°
we got the Wind at West, the next day it fixd at N.N.W.
and increased to a fresh gale, with which we steered directly
for the Lizard   and on

Sat: 29.
We made the Land about Plymouth, Matter Church,
at 5 o'clock in the After-noon, bore N.10.W. distant
4 leagues; this bearing and distance shew'd that the
error of M.' Kendals Watch, in Longitude was only 7'..45,
which was too far to the West ——————

Jam.s Cook

XXIV. Larcum Kendall's facsimile of Harrison's No. 4 Chronometer, and Captain Cook's testimony to its accuracy at the close of the Journal of his second voyage.

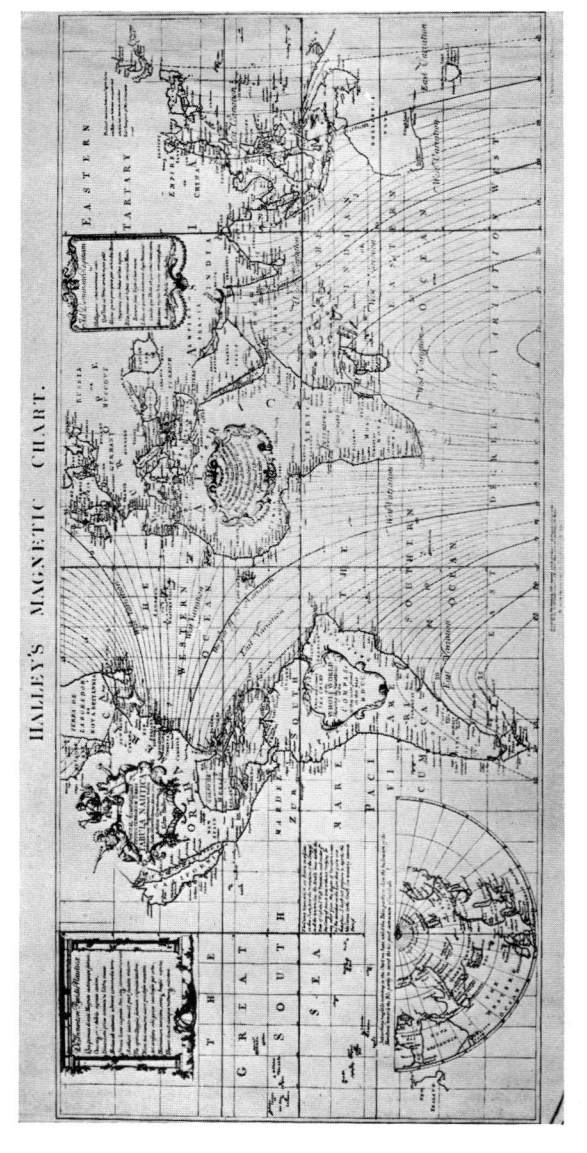

XXV. Halley's magnetic variation chart of the world.

admiral, Sir Cloudesley Shovell. And in 1713 two mathematicians wrote to the Press declaring that this shocking waste of life had been due to the seaman's ignorance of his longitude, and was preventable. The two were William Whiston, recently expelled from the Lucasian Chair at Cambridge on a charge of heresy, and Humfrey Ditton who had been given a special mathematical teaching post at Christ's Hospital through Isaac Newton's influence. Between them they believed that they had discovered a method for the longitude that cut through all the old difficulties; it was 'easy to be understood and practis'd by ordinary seamen, without the necessity of any puzzling calculations in astronomy'. Public opinion was stirred, a Bill was introduced into Parliament, and in 1714 an Act was passed 'For providing a Publick Reward for such Person or Persons as shall discover the Longitude at Sea'. A Board of Longitude was established, the Commissioners including both seamen and scholars, and the maximum prize was fixed at £20,000, while up to £2000 could be granted to help experimental work deemed promising. The method proposed by Whiston and Ditton, which depended on simultaneous light and sound signals sent up from anchored hulks lying in precisely charted positions, was not a practicable one, and though commended by Newton was held up to ridicule by hard-headed men. But the offer of a prize brought in many even less useful suggestions. The method was required to be one which would serve on a voyage to the West Indies and back with a total error of less than 30' or 2 minutes of time, and this ruled out the pamphleteer who proposed that time-signals should be flashed on the clouds from lighthouses. Several people thought that they qualified for the reward by suggesting that a correct timepiece should be taken to sea, although one or two went further and claimed to have designed such a clock or watch. An instrument-maker put forward a 'marmeter' or improved mechanical log, while a tutor suggested constructing a universal tide-table and measuring the rise and fall of the tide on the high seas by means of a portable barometer! A number of those who promptly published pamphlets claiming to have methods for the longitude did not reveal what these were, and in fact there was no suggestion made of any substance and the excitement died down. Nevertheless the Act attracted great attention overseas, and particularly in France. There the two Prix Rouillé were founded in 1715 for scientific work presented to the Academy, the second to have particular regard to navigation and commerce. This second prize stimulated a good deal of work on the longitude, particularly among the clock-makers.

For ten or fifteen years after the passing of the Act of Parliament of 1714 there seems no progress to report in the history of navigation. The training of boys at the Christ's Hospital mathematical school remained under the charge of James Hodgson, F.R.S., a sound mathematician, right down to 1748 when he was nearly eighty. John Flamsteed died in 1719, to be succeeded as Astronomer Royal by the now elderly Edmund Halley, already long past his wonderful creative period. He held the office until he was eighty-six. The great star catalogue, which was Flamsteed's monument, was published in its definitive form in 1722. With the passing or the ageing of the giants of its young days the Royal Society's activity and influence waned for a period, but there was a wonderful flowering of instrument-makers in London. Such names as George Graham, Jonathan Sisson, Thomas Wright, John Adams, Thomas Heath, are representative of a much longer list of men who are remembered and admired today for their beautiful and accurate workmanship. The Hanoverian kings, following continental precedent, extended royal patronage to such instrument-makers, and finely ornamented orreries, globes, microscopes, magic-lanterns and optical toys had their place in a gentleman's library. It was natural therefore that a descriptive work on mathematical instruments should be called for, and a young man who had grown up in the household of the Duke of Argyll was persuaded to translate a new French work, *La Construction et l'Usage des Instrumens de Mathématique*, by Nicholas Bion, instrument-maker to the French king.

The young man was Edmund Stone, the gardener's son, who had taken a passion for mathematics as a boy and been encouraged by the duke. 'Mathematics are now become', he wrote in his Preface in 1723, 'a popular Study, and make a part in the Education of every Gentleman', and he goes on to say: 'Mathematical Instruments are the means by which these Sciences [i.e. the mathematical sciences] are rendered useful in the Affairs of Life.

By their Assistance it is, that subtill and abstract Speculation is reduced into Act.' The whole volume is divided into eight books, of which the seventh deals with instruments used at sea, and at the end of each book the English translator adds notes on such English instruments as were not fully described by Bion. No such supplement was, however, needed for Book VII, for the same practices held on either side of the Channel, and the really essential instruments, apart from the sea-compass, were for both nations the azimuth compass and the 'English quadrant' as the French called the back-staff. In addition there was Mercator's chart for pricking the course, and the cross-ruled 'sinical quadrant', essentially a fourth part of Gemma Frisius's 'geometrical square', besides the various 'traverse boards', all of which provided a graphical means of finding distance and departure. The more modern of these last instruments carried a marginal scale of meridional parts, and the user was taught to find the middle latitude of a traverse before turning the east–westing into leagues. The Mercator's charts, says the French author, 'are commonly esteemed the best; for by the experience of several Ages, it is found that Seamen ought to have very simple Charts, wherein the Meridians, Parallels and Rhumb-Lines may be represented by straight Lines, that so they may prick down their Courses easily'. But all these things were more than a century old, as was the azimuth compass (Plate XX), looking much as it had been designed by the Elizabethans, William Barlow and Robert Dudley.

Nor had the 'English quadrant' undergone any special improvement since John Davis's day, save for the introduction of a lens instead of a shadow to cast the Sun's image on the horizon slit. The great advantage and popularity of this instrument lay not so much in the fact that the observer could turn his back on the Sun, as in the relief it gave him from the well-nigh impossible task demanded by the cross-staff of simultaneously holding two separate observations—star and horizon—correctly (Plate XIV). Once he had the Sun's rays passing at an approximately correct angle through the lens on the upper quadrant he could now concentrate his whole attention on getting the bright spot and the horizon precisely aligned by means of the sight on the lower quadrant. The scale on this lower quadrant was to single degrees subdivided by a diagonal scale, but although sailors were quite satisfied with the back-staff, since much greater errors of reckoning sprang from other causes, it was but a crude instrument from an astronomer's standpoint. Nor was it free, of course, from all the inconvenience and errors which arose from the motion of the ship.

This last disadvantage was dramatically overcome by a new octant displayed to the Royal Society in 1731 by a Fellow, John Hadley, who had long been interested in the making of improved optical instruments. In particular he had worked on the reflecting telescope with J. T. Desaguliers, a prominent mathematician of the day, and had shown a new model, which he termed 'catadioptric', to the Society in 1723. His octant (Plate XXII) made use of Newton's simple principle of reflecting the heavenly body down to the point on a second mirror to which the horizon had already been aligned. The second (fixed) mirror was half quicksilvered, half plain, and when the mirror on the index arm stood at zero the horizon (i.e. the edge of the sea) was brought to such a position that the line seen directly through the glass matched exactly the mirrored line. The angle through which the index arm must be turned to bring down the heavenly body to the horizon is half the angle of its elevation. Hence an arc of 45° is marked on the scale of the limb as 90°. When the instrument was used to observe the Sun, the eye was protected by coloured glass and, of course, correction had to be made for the semi-diameter of the luminary, whether its upper or lower limb was brought to the horizon. The octant had a length of about 20 in., and could be firmly grasped in one hand while the other moved the index arm. It was necessary to hold it as nearly as possible in the vertical plane. What Hadley claimed was the chief advantage of the instrument was that whereas in taking an altitude by all others a certain posture was demanded which the motion of the ship constantly disturbed,

> 'with this Instrument, though the Ship rolls ever so much, provided the Instrument be kept in or near an upright Posture, though it be leaned forward or backward therein, yet the Image of any Object, when once brought by sliding the Index [alidade] to appear on the Edge of the Sea, will there remain absolutely immoveable as long as the Index continues in the same Place, without being stirred, and the Observer has the same Advantage of making the Observation as if he took it in smooth Water, and the Instrument was held still without Motion.'

The value of the instrument was at once apparent and the Lords Commissioners of the Admiralty ordered that it should be tested at sea. This was done in August 1732 in the yacht *Chatham*, John Hadley being assisted by his brother George and the astronomer Bradley. The readings were found accurate to 2', and in many cases to 1', and the success of Hadley's octant (sextant as, when the scale was extended, it came to be called) was assured, although a large proportion of sailors clung to the old back-staff. Whether Hadley knew, or had been influenced by Hooke's or Newton's earlier work with mirrors is immaterial, but certainly their work was well known to Halley, and, when the latter died in 1742, Newton's description of his reflecting quadrant was found among his papers and read to the Royal Society. The actual instrument was in the possession of the instrument-maker, Thomas Heath, and had been exhibited in his shop window, perhaps before the paper was found. In 1739 a pamphlet was published in France called *Le Nouveau Quartier Anglois*, and the engraving shows a simple form of sextant beside a back-staff. The writer was Lieutenant d'Après de Mannevillette, an officer of the French East India Company, who was later to become a notable hydrographer, and the quadrant was advertised as made by le Maire, fils, an instrument-maker on the Quay de l'Horloge du Palais. There was no mention, however, of Newton's name, but the general question of his priority over Hadley was raised by Edmund Stone, who prepared a second edition of his book on instruments in 1758. He then added an Appendix in which was a chapter 'Of some more modern Instruments used at Sea', and there he stated categorically: 'The first of these instruments . . . was invented long ago by Sir Isaac Newton.' But in fact the idea of bringing two objects together by reflection was very much 'in the air' at the time when Hadley read his paper to the Royal Society.

In America, a Philadelphian glass-worker, Thomas Godfrey, had independently devised an instrument similar to Hadley's, which only reached London in 1732, while in France somewhat the same sort of instrument (with a single mirror) was designed by Fouchy. Another English inventor was Caleb Smith, whose instrument was made and sold by Thomas Heath, together with a descriptive pamphlet entitled: 'The Use of the new Instrument, or Sea Quadrant, for taking Altitudes of the Sun, Moon and Stars, from the visible Horizon by an Observation either forwards or backwards, as well as any other angular Distances, without Impediment or Interruption from the Ship's Motion, whereby the Latitude at Sea may be obtained with greater Certainty and more frequently, than by any other Instrument commonly used for that purpose.' In this instrument prisms were substituted for mirrors, but as it proved none was so simple and convenient in use as Hadley's. A French observer, writing in 1755, said he had seen fifty of Hadley's for every one of Smith's in use on French ships, and Bouguer, writing his *Traité de la Navigation* two years earlier, had said: 'On a imaginé en Angleterre un nouvel Instrument incomparablement plus parfait que ceux dont nous venons de parler.' He had just been describing the ordinary cross-staff and back-staff, and the incomparably more perfect instrument was Hadley's.

Bouguer was one of the French Academicians sent to Peru to measure the arc of the meridian at the equator in 1735, and they had taken one of the new octants with them. He himself had spent his life in the world of navigational theory and practice. His father had been one of the more notable sea-port professors, while the son had held that office first at Croisic, and then at Le Havre where he won fame by his successful prize essays on observations at sea. His brother, too, followed the same profession, so that altogether his verdict on the octant could claim to be soundly based. The best instruments came from London, and those of Edward Nairne were particularly praised. They had a frame of oak, a limb of ivory and an index arm of copper. A telescopic sight was considered unnecessary in view of the continuing errors inevitable in longitude estimation, but some instruments were furnished with a vernier, and with a bubble-level which indicated whether they were being held correctly.

Difficulties, however, were far from over. It was still easier to make a good plane mirror of polished metal than of silvered glass, but the metal readily tarnished. Then, too, the mirrors required a very precise adjustment of their positions on the instrument, since the exactness of the reading depended on the angle between them being true. Nevertheless, the 'sextant' was a triumph for the sailor, placing him no longer at the mercy of the waves, and even John Robertson, in the heavy textbook he wrote for the Christ's Hospital boys (he taught there 1748–55), spoke of it as 'this most excellent instrument'. To d'Après de Mannevillette, the 'quartier Anglois' appeared also to offer fresh hope of obtaining the longitude by the method of lunar distances, the type of observation that Newton had had in mind, and he made use of it for this purpose in his hydrographical surveys from 1749 onwards. At first, owing to the imperfections of his instrument, the results were disappointing, but he had a new octant made by Morgan of London which was more successful, and claimed that, owing to his extensive practice with it, he could count on a reading correct to within 2′ or at most 3′.

Meanwhile, however, so far as the longitude was concerned advances were being made in quite a different direction, that of the marine clock or chronometer. These advances were at first very slow, perhaps because astronomers in general had made up their minds that the only way of success lay through observations of the Moon—not only the eminent Halley, for example, but a later Astronomer Royal, Nevil Maskeleyne (founder of the *Nautical Almanac*) held to this opinion. The chronometer, in fact, came from the instrument-makers, and not from the mathematicians, and all the work of importance was done in France and England, the two great naval powers of the century. Exceptionally, however, the first award went to a Dutchman, Massy, who in 1720 won the first Rouillé prize, for which the subject set was the regulation of the pendulum at sea. Massy's solution, although it passed muster with the Academicians, appears to have been a complicated and cumbersome device which could not have been brought into practical use, and indeed in 1725 the Academy asked for research to be made into improving the water-clock and the sand-glass, which sounds like a counsel of despair. Nevertheless, there was more sound sense among the sailors, and one of them, Radouay, in his *Remarques sur la Navigation*, published in 1727, reported that in the course of a voyage to Newfoundland in 1722 he had been able to make substantial corrections in his dead reckoning with the help of some good watches that happened to be aboard.

The first actual progress made was by an Englishman named Sully, a clock-maker who had settled in Paris in 1712. He made several improvements in clock-work, especially in lessening friction, and a very accurate pendulum of his was tested briefly at sea in 1726, the results, however, not being conclusive. He had, too, advice from Sir Isaac Newton and Leibniz, and consultations with the famous George Graham, F.R.S., notable for his instruments of precision and inventor of the dead-beat escapement. But Sully's importance rests perhaps on the training which he and his establishment gave to Pierre Le Roy, son of his fellow-workman Julian Le Roy. The younger Le Roy was to make a French chronometer, although he was forestalled by many years by the English clock-maker, John Harrison.

The son of a Yorkshire carpenter, Harrison became interested in mathematics as a very young man, and did some surveying, but he settled down in London as a clock-maker, and not until 1729, when he was approaching middle life, did he attempt a marine clock. His pendulum clocks, however, which were so perfect that they did not gain or lose a second in a month, had attracted attention some years earlier. After six years' work his first chronometer was completed in 1735 and in 1736 was tested on a voyage to Lisbon and back when the ship's captain paid tribute to its usefulness (Plate XXIII). It enabled him to correct an error of $1\frac{1}{2}°$ in his dead reckoning at the entry to the Channel. In the following year the Commissioners for Longitude exercised their powers to make Harrison a grant of money towards further research, and he completed two more clocks in 1739 and 1741. In 1749 he was awarded the Copley Medal of the Royal Society, the president, Folkes, delivering an address on the labours and successes which had earned it. But the commissioners (who included the Astronomer Royal, and the holders of mathematical chairs at Oxford and Cambridge) were not yet satisfied. Harrison, now an old man, embarked therefore on a watch-type timepiece, which was not completed in 1761. He then immediately demanded that the trials detailed under the Act of 1714 should be carried out, these including a voyage to the West Indies and back. He requested, also, that his son William should be present at the trials, and the arrangements were promptly made. The younger Harrison was ordered to report at Portsmouth where the ship *Deptford* was about to leave to convey a Governor to Jamaica.

Chronometer No. 4 was officially timed at Portsmouth, and a report sent to the Admiralty. It was then put aboard under lock and key. An astronomer, Robinson, accompanied the party, who was to determine the local time in Jamaica.

Within three weeks of embarkation an unexpected test took place. The ship was approaching Madeira, and the chronometer showed a longitude of 15° 19′ west of Portsmouth. But the pilots made it only 13° 50′, and all on board (save Harrison) immediately concluded that the instrument was a failure. The more so as the voyage was so familiar, and it was the general experience that dead reckoning was likely to set a man too far west, and not too far east of his true position. The pilots demanded that they should change course in order (as they supposed) to reach Madeira, where they were to take on supplies, and the captain was ready to agree, but Harrison begged him to put off the change until the next morning, when, if the chronometer was correct, Porto Santo would be sighted. Fortunately Captain Digges consented, and the island duly came into sight. The chronometer had triumphed, at least in the eyes of the crew. Port Royal, Jamaica, was reached in January 1762, and the longitude of this place had been comparatively recently correctly fixed by eclipses of the Moon and by a transit of Mercury across the Sun. The figure expressed as an hour angle was 5 hours 16 minutes 23 seconds, and the watch differed from this by only 5·1 seconds. The time between its setting at Portsmouth and this comparison was 81 days, so that here indeed was a success for Harrison. The vessels of an English fleet that had followed the *Deptford* were in error by as much as 5°, or 20 minutes of time. On the return journey in the *Merlin* very bad weather was experienced, and Harrison had had the chronometer fixed solidly and not hung in gimbals, but the total error after a lapse of 147 days was under two minutes, 1 minute 54½ seconds to be exact. Old Harrison could claim that he had fulfilled the conditions of the Act, which laid down that for the full reward the longitude must be determined within half a degree. But not all members of the Board of Longitude were satisfied, and instead of £20,000 he was granted only £2500, although in the following year Parliament voted him a further £5000. A second trial was demanded and in 1764 the younger Harrison took the chronometer to Barbados and back.

On this occasion the instrument lost only 15 seconds in 156 days taking into account the maker's table of corrections for temperature change. Even ignoring these it lost only 54 seconds. On February 9th, 1765, the Board declared unanimously therefore that the conditions of the Act had been fulfilled. Yet they only brought the old man's reward up to £10,000. And no doubt they were right in their contention that it was implicit in the Act (although not actually stated) that the reward was to be given for a method fully explained, and so generally available. They demanded therefore that Harrison should explain the construction of his chronometer to the satisfaction of the commissioners and of a nominated panel of mathematicians and horologers. It was said by critics that the resulting brochure he produced gave no clear explanation of his methods, and this may well have been the case, for he was a craftsman rather than a savant. But he was acclaimed in France in the *Connoissance des Temps* for 1765. 'Il est juste', wrote the editor, 'que Harrison jouisse en la France de la gloire dont on le juge digne dans sa patrie'. Meanwhile Maskeleyne, the Astronomer Royal, took the chronometer to Greenwich observatory and studied it minutely day by day in the most exacting and critical fashion that he could contrive. It had faults, he finally declared, but nevertheless was an invention which would be very advantageous to navigation. Subsequently the Board of Longitudes commissioned another clock-maker, Larkum Kendall, to make a facsimile of Harrison's fourth chronometer (Plate XXIV). This model was carried by Captain Cook on board the *Resolution* in 1772, and proved of immense value. Harrison was completely vindicated and was given the balance of the award in 1775, the year before he died. Yet it is surprising how slight had been the impact of his work in quarters where it might have been expected to be eagerly welcomed. Edmund Stone in 1758 knew only of his clocks vaguely by hearsay, while John Robertson, the nautical instructor, merely remarked that chronometers would have to become very much cheaper than they were before they could be considered as a method for finding the longitude. Which was true.

Nevertheless the battle was really won, and priority rested with England, although the work that was being carried out in France did not lag very far behind. Pierre le Roy had two chronometers ready for testing in 1766, for the subject set for the Rouillé prize for 1767 had been 'the best way of determining time at sea'. But he had a rival, the Swiss clock-maker Ferdinand Berthoud, who had been settled in France since 1745, and had gained public attention by his *Essai sur l'Horlogerie* published in 1763. Berthoud's first clock to be tested at sea did not give very good results and he was twice sent to London to try and learn something from Harrison, though without success, but by 1766 he too had new instruments ready for testing. Le Roy's two clocks were first of all taken to sea at the expense of a private patron, and the tests were inconclusive, but sufficiently promising to lead to an official trans-Atlantic test in which the younger Cassini took part. As a result Le Roy was given the deferred prize, and meanwhile the distinguished scientist Fleurieu sailed with Berthoud and his clocks. The period of controversy and rivalry between the two makers lasted through the century, but it was of no consequence, for the totality of improvement of clockwork during the struggle meant that the longitude was truly solved. And the longitude being solved, there was no longer justification for any checks on the improvement of the sextant, or for any carelessness in the determination of latitude and local time. Maskeleyne's first Nautical Almanac appeared in 1767, and with Hadley and Harrison what may be termed the pre-scientific age of navigation was brought to a close. Landmark or no landmark, the sailor knew precisely where he was—or had the means to know. He did indeed at long last possess the Haven-finding Art.

APPENDIX

Navigation in Medieval China

In tracing the history of navigation in the West we have examined the origins in antiquity and the impetus to mathematical methods provided by the renaissance of classical learning in the age of the great discoveries. In this connection we noted the importance of European contacts with the Arab world. While there were few sustained contacts to record with far-away China before the travels of Marco Polo, a transfer of skills may well have taken place, for the art of navigation had long flourished there, anticipating most of the advances made in Europe before the beginning of the seventeenth century. From primitive origins, no doubt similar in character to those in the West, its development had included the use at sea of the magnetic compass and the determination of latitudes from the elevation of selected stars. By these means the great Chinese ships of the fifteenth century were piloted across the southern oceans to the East Indies, India, Arabia and the eastern coasts of Africa. Had the Portuguese reached these waters a generation or so earlier than they did, they would have been confronted by a naval power and navigation techniques more sophisticated than their own. As it was, Chinese naval power had waned before the first European contacts were made by sea, and their long established arts of navigation were eclipsed by the more recent advances of the European mathematical practitioners, transferred through the hands of Portuguese and Dutch sailors. To provide a brief outline of the development of Chinese navigation, the present chapter has been distilled by Frank George, with the permission of the author, from the section on Nautical Technology in Volume IV (Part 3) of Joseph Needham's great work, *Science and Civilisation in China*. (Acknowledgements are gratefully made to the Cambridge University Press.) Documentation of evidence, and the characters for Chinese names and terms, will be found therein.

It is hardly to be questioned that from the earliest times when Chinese shipmasters sailed their vessels out of sight of landmarks they steered by the stars and the Sun. Chang Hêng was probably referring to their starcraft when he wrote in his *Ling Hsien* (A.D. 118): 'There are in all 2500 (greater) stars, not including those which the sea people (*hai jen*) observe.' This could equally well have been translated 'sailors', and raises the ghost of a literature long lost and now hard to interpret, but perhaps highly relevant. The books that we find in the Han bibliography of the first century A.D. include astrological and astronomical manuals for the *hai jen*. In the astronomical monograph of the *Thang Shu* it says that 'in the twelfth year of the Khai-Yuan reign-period (A.D. 724) the Astronomer-Royal was instructed by decree to proceed to Chiao-chou (Hanoi) to measure Sun-shadow lengths. There, while at sea (*hai chung*) in the eighth month, looking southwards, they observed the remarkably high altitude of Canopus (*Lao jen*). Below it there were numerous stars brightly shining, including many large ones, but these had not then been recorded on the celestial maps, and their names were not known.'

Probably we shall not go far wrong if we identify the *hai chung* (sea-going) corpus as the work of 'magicians' of the Warring States period and Early Han who lived along the coasts of Chhi and Yen, the 'mathematical practitioners' of the earliest stages of Chinese navigation. Their skills were doubtless undifferentiated, and it would be impossible to disentangle in them the components which today we should call astrology, astronomy, stellar navigation, weather-prediction, and the lore of winds, currents and landfalls; all the more so since (as in the work of Dee, Hartfill, Goad, Gadbury, and many others) these elements were still wholly confused down to the end of the seventeenth century in Europe. At all events we can now form some idea what kind of men those 'sea-going magician-technicians' were.

Chinese pilots in the period of primitive navigation certainly made use also of all these ancient aids. But it was they who brought this period to an end by being the first to employ the magnetic compass at sea. This great revolution in the sailor's art, which ushered in the era of quantitative navigation, is solidly attested for Chinese ships by A.D. 1090, just about a century before its initial appearance in the West. Our first text which shows this also mentions astronomical navigation and soundings, together with the study of sea-bottom samples. Two further accounts in the twelfth century follow before the first European mention, each emphasising the value of the compass on nights of cloud and storm. It may be significant that one of these specifically says that at night the pilots steer by the stars and the Great Bear. This may mean no more than was intended by Aratus in the third century B.C., when he noted that the Greeks steered by Helice (the Great Bear), but on the other hand it might imply the beginnings of altitude measurements. The complement of Aratus is doubtless the passage in the *Huai Nan Tzu* book of about 120 B.C., where we read: 'Those at sea who become confused and cannot distinguish east from west, orient themselves as soon as they see the Pole Star.' The exact date at which the magnetic compass first became the mariner's compass, after a long career ashore with the geomancers, is not known, but some time in the ninth or tenth century A.D. would be a very probable guess. Before the end of the thirteenth century (Marco Polo's time) we have compass bearings recorded in print, and in the following century, before the end of the Yuan dynasty, compilations of these began to be produced.

In all probability from the beginning of its use at sea, the Chinese compass was a magnetic needle floating on water in a small cup. A thousand years earlier, the first and oldest compass had been a spoon-shaped object of lodestone rotating on a bronze plate. At some intervening period the frictional drag of the spoon on the plate had been overcome by inserting the lodestone in a piece of wood with pointed ends, which could be floated, or balanced, upon an upward-projecting pin. The dry-pivot compass had thus been invented; but although these primitive arrangements seem still to have been used as late as the thirteenth century, Chinese sailors did not (so far as we know) employ them. For at some time between the first and sixth centuries A.D. the discovery had been made that the directive property of the lodestone could be transferred by induction to the small pieces of iron or steel which the lodestone attracted, and that these also could be made to float upon the surface of water by suitable devices. The earliest extant description of a floating compass of this kind dates from just before 1044 A.D. and involves a thin sheet of magnetised iron with upturned edges cut into the shape of a fish. To floating compasses of one kind or another Chinese navigators remained faithful for nearly a millennium. We have detailed accounts of their use from the fifteenth century. But in the sixteenth there came Dutch influence, mediated in part through the Japanese, as a result of which the dry-pivoted needle and then the compass-card (doubtless an Italian invention) were adopted on Chinese vessels. The Chinese compass-makers, however, employed a very delicate form of suspension which automatically compensated for variations of dip, and still impressed western observers as late as the beginning of the nineteenth century.

The remarkable series of maritime expeditions led by the admiral Chêng Ho between 1400 and 1433 must always remain a focal point in the history of Chinese navigational technique. By great good fortune, certain maps of portulan character dating from about this period, which trace the routes followed by these and other Chinese ships and convoys, have been preserved intact. The term portulan is used here only loosely and non-technically, for the Chinese sailing charts had no intersecting rhumb lines and no grid. Early in the seventeenth century they were printed as the last chapter of an important treatise on military and naval technology, the *Wu Pei Chih*. These charts are extremely distorted but schematic, and ships' courses are drawn across their oceans like the tracks in the brochures issued by modern steamship companies. The lines of travel are accompanied by legends giving detailed compass bearings, with distances in numbers of watches (*kêng* or *ching*), and notes of most of the coastal features which could be important in navigation. The bearings are always given in the form *hsing ting wei chen* (sail with the needle between *ting* and *wei* azimuth points, i.e. south-south-west), or *yung kêng shen chen* (use the needle pointing between *kêng* and *shen*, i.e. west-south-west), while *tan khun* (bear on red, or single *khun*) meant sailing with the needle pointing directly to *khun*, i.e. within $3\frac{3}{4}°$ on each side of south-west. Thus the general formula is: 'sail on such a course for so many watches.' The notes include indications of half-tide rocks and shoals as well as ports and havens. Routes are given for inner and outer passages of islands, sometimes with preferences if outward or homeward-bound. Much attention has been devoted by modern scholars to the accuracy of these diagrams and descriptions and to the identification of the place-names, and a high opinion has been formed of the knowledge and precision of these Chinese navigators' records. Some idea of the skill of the pilots may be gained by the fact that in circumnavigating Malaya they laid their course through the present Singapore Main Strait, which was not discovered (or at least not used) by the Portuguese until they had been in these waters for more than a hundred years.

The interest of the last chapter of the *Wu Pei Chih*, however, is not exhausted by these schematic charts. Four instructive navigational diagrams are given, summarising the star positions to be maintained during as many regular voyages. In the diagram of pilots' directions for that between Ceylon (Hsi-lan Shan) and Sumatra (Su-mên-ta-la, Kuala Pase, modern Samudra), for instance, a reading of the notes concerning the 'guiding stars' (*chhien hsing*) which are distributed round the central picture will lead us into the heart of the matter.

'[*Above*] The Pole Star (*Pei chhen*) to be 1 *chih* above the horizon, and the Imperial baldachin (*Hua kai*) 8 *chih*.

[*To the left*] In the north-west the *Pu ssu* stars to be 4 *chih* above the horizon; and the same in the south-west.

[*Below*] The Frame (or Bone) of the Lantern (i.e. the Southern Cross) to be 14½ *chih* above the horizon. The twin stars of the 'Southern gate' (*Nan mên*) to be level at 15 *chih*.

[*To the right*] In the north-east the Weaving girl (*Chih nü*) to be 11 *chih* above the horizon.'

The explanation of all this lies in the fact that for measuring the altitudes of the Pole and other stars the pilots did not use the degrees of the astronomers, but rather another graduation in finger-breadths (*chih*), each of which was divided possibly into 8, but more probably into 4, parts (*chio*). Moreover, for this voyage the Pole Star was very low on the horizon or invisible, and it was therefore necessary to substitute for it a circumpolar markpoint, the *Hua kai* constellation. Altitude on this would be measured each night when it culminated, and the altitudes of all the other guiding stars taken presumably at the same time, and to establish the time of culmination. The *Wu Pei Chih* charts give *Hua kai* altitudes for a number of places.

Once we realise that the navigators of the China Seas and the Indian Ocean depended quite as much on polar altitudes as the Portuguese came to do towards the end of the fifteenth century, a host of fascinating questions arise. Unfortunately we know as yet neither exactly how far back this quantitative oceanic navigation went in Eastern waters, nor how far the Europeans of the Atlantic border were influenced by it during the exploration of the West African coast. Certain it is that when the Portuguese showed him their astrolabes and quadrants in the summer of 1498 Ibn Majid, the pilot who guided Vasco da Gama from Malindi to Calicut, was not in the least surprised, saying that the Arabs had similar instruments; but the Portuguese were very astonished that he was not surprised. Moreover, there are a number of points at which we may suspect East Asian influence on Europe, or where at least we have to grant considerable East Asian priority.

First, it is clear that the Chinese navigators of Chêng Ho's time, besides their compass-bearings, knew the method of finding and running down the latitude. In the *Hsi-Yang Chhao Kung Tien Lu,* for example, there is talk of a voyage from Bengal (probably Chittagong) to Male in the Maldive Islands, by way of Ceylon, and the polar elevations are given for every stage of the journey. Thus a certain Ceylonese mountain will be sighted when the altitude has sunk to 1 *chih* 3 *chio*. We are still uncertain as to the instruments the Chinese used. By 1400 quadrants would be quite possible—the armillary sphere had had a long and elaborate history in China, and some such apparatus had been used overseas as far back as the beginning of the eighth century, when I-Hsing's meridian arc survey teams took altitude measurements from Indo-China to Mongolia. That was the time, too, when a southern hemisphere astrographic expedition had been sent to map the constellations to within about 20° of the antarctic pole. Astrolabes as known in the West would not be in the picture, but a simplified armillary ring with swinging alidades or the characteristically Chinese sighting-tube may well have been used ashore. Even more probable, seemingly, would be the simpler types of cross-staff, for elsewhere evidence has been given that Jacob's Staff was known in China and used by surveyors three centuries before the description of Levi ben Gerson, i.e. by 1086 rather than 1321. This would also be more in line with the practice of the Arab and Indian pilots.

The problem of maritime charts is also very obscure. That they existed is implicit in many Chinese texts, but the only ones which have survived are the schematic diagrams, almost like the 'Peutinger Tables', preserved in the Wu Pei Chih. Nevertheless, the tradition of quantitative cartography was much stronger in China than it was in Europe, so that already by. 1137 a superb map on a scale of 100 li to the division could be produced, and there is little reason to think that the much larger map of Chia Tan on a similar scale in 801 was any less good. Indeed, the principle of the rectangular grid went back to Phei Hsiu in the third century A.D. It is thus of particular interest that in the work of Shen Kua, late in the eleventh century, we do have a hint that the grid was combined with compass-bearing rhumb lines, just as occurred two or three centuries later in the Mediterranean; but his work was terrestrial, not nautical, and it did not survive. Lastly, the projection of Gerard Mercator in 1569 was a great advance, but he never knew that he had been preceded by Su Sung five centuries earlier in a celestial atlas, in which the hour-circles between the hsiu (lunar mansions) formed the meridians, with the stars marked in a quasi-orthomorphic cylindrical projection on each side of the equator according to their north polar distances. With such a brilliant background we must hope that archaeological discoveries will yet reveal what charts were used by the master-mariners of the Sung, Yuan and Ming.

In Europe the magnetic compass, the portulan chart, the sand-glass and the marteloio formed a closely connected knot of complementary techniques. Little can be said of traverse tables (marteloio), which have not so far been recognised in Chinese rutters. It has been thought that the sand-glass was not known or used on Chinese ships until the end of the sixteenth century, when they acquired it from the Dutch or the Portuguese, but an important development in Chinese mechanical clockwork in about 1370 substituted sand for water in clocks of the classic scoop-wheel type, and there is no reason why such clocks should not have been carried (as the older water-wheel ones could not have been) on the great ships of Chêng Ho's fleet. In any case it is clear that time-keeping by sand-flow was very much in the minds of the Chinese at that time. It is necessary therefore to re-examine the Western traditions which make the sand-glass begin with Liutprand of Cremona in the tenth century, and to reconsider the suspicion of Speckhart, long ignored, that the hour-glass came to Europe from the East. Since nautical watches (*kêng*) are mentioned (or implicit) in many descriptions of Chinese navigation from the beginning of the twelfth century onwards, the measurement of such units must have necessitated a timepiece, and no form of water clepsydra would be imaginable at sea. But against the use of the hour-glass there is a serious argument and another way out. The 'sailor's dyoll' implies blown glass, and the glass-blowing art appears to be wholly European and Western, though by no means glass-making itself. On the other hand time-keeping by burning incense in sticks or trails is a practice which goes far back into China's Middle Ages, and it would have been very easy to measure time approximately enough with 'joss-sticks' kept alight in the ship's shrine, where the compass lay also. The incense-stick was so characteristic of Chinese religion and culture that it may have been difficult for it to spread to mariners of other cultures, even though they might have found it very useful.

Let us now return to the altitude measurements in *chih* and *chio*. The remarkable feature of this system is that it was practically identical with that in use among the Arabic shipmasters of the Indian Ocean, who expressed altitudes in *isba* (the finger-breadth or inch) equalling 1°36′25″, and its eighth part, the *zam*. This must have been in full employment at the time of Chêng Ho's voyages. Moreover, when the measurements in the *Muhit* and its sources are compared with those in the *Wu Pei Chih*, they are found to be, generally speaking, in good agreement. The chief difference between the Arab and the Chinese systems seems to be that when a 'substitute' polar mark-point was desired in equatorial latitudes the Arabs chose the classical 'Guards' (β and γ *Ursae Minoris*), which they called *al-Farkadain* (the Calves), while the Chinese chose *Hua kai,* the declinations being very similar, but the right ascensions almost exactly 180° (12 hours) apart. Surely the only likely explanation for this is that the Arabs and the Chinese were at some time or other accustomed to sail in these southerly latitudes at different times of the year. The practice which suited each group would then have become a convention. Perhaps the Chinese were avoiding the typhoon season and the Arabs the monsoon. Seen from the southern tropics *Hua kai* culminates about midnight toward the beginning of November, and would be usable between August and February; *Ursa Minor* culminates about midnight in early May, and would be usable between February and August. Both Arabs and Chinese took an elevation of 1 finger-breadth as the point at which it was no longer safe to trust to Pole Star measurements; they then changed to a circumpolar mark-point, the Arabs taking 8 finger-breadths of *al-Farkadain* and the Chinese 8 finger-breadths of *Hua kai* as equivalent to 1 finger-breadth of Pole Star elevation.

The question may be raised as to the mutual influence of the Arabic and Chinese navigators, but at present we hardly know enough to answer it. They had certainly been in contact for many centuries before 1400 A.D. Measurements of altitude were particularly prominent in all Arabic astronomy, but on the other hand circumpolar mark-points for invisible stars were rather characteristically Chinese. Again, *chih* and *chio* measurements are not common in early printed Chinese texts, but that does not mean that they were not in widespread use by pilots, whose rutters were generally hand-written. Moreover, we can find possible Chinese mentions of surveyors' measurements in finger-breadths as far back as 260 A.D., a good deal earlier than anything similar in Arabic culture. But of course the system of finger-breadth units for altitudes could easily have arisen independently in the Arabic and Chinese culture areas.

What instruments were used by the Chinese pilots of the Yuan and Ming for taking star altitudes was long a puzzle, but of those employed among the Arab sailors a good deal is known; they were all forms of the cross-staff or Jacob's Staff, including the tablet or *kamal*. It remains extremely probable, therefore, that the Chinese pilots of the fifteenth century used some form of cross-staff. That they used a version of the *kamal* has now been proved by Yen Tun-Chieh's brilliant interpretation of a passage in the *Chieh An Lao Jen Man Pi*, by Li Hsü, who lived from 1505 to 1592:

'The set of "guiding star stretch-boards" (*chhien hsing pan*) of Ma Huai-Tê of Suchow, has twelve plates in all, made of ebony, ranging gradually from small to large. The largest is more than seven inches square [lit. long]. They are labelled "one *chih*", "two *chih*" etc., up to "twelve *chih*", all marked in fine script upon them; and they differ regularly just as a foot is divided into inches. There is also one ivory piece, two inches square [lit. long], and cut off at the corners so that it indicates half a *chih* (i.e. two *chio*), half a *chio*, one *chio* and three *chio*. This may be turned on one side or another facing you (in conjunction with one of the larger plates), and these lengths must be the measurements required for right-angle triangle calculations according to the methods of the *Chou Pei* (*Suan Ching*, Arithmetical Classic of the Gnomon and the Circular Paths of Heaven).'

Evidently we have here a set of standard ebony tablets held at a fixed distance from the eye, not the single one with its knotted string that constituted the typical *kamal*; plus the interesting addition of a 'fine adjustment' in the shape of an ivory tablet with corners truncated to small standard edge lengths, held up at the same time to allow the measurement of fractions of a *chih*. Yen Tun-Chieh's calculations showed that the series of tablets described corresponds to a range of from 1°36' to 18°56' of altitude, with an average difference of 1°34'30" representing the *chih*. It is equally clear that the Chinese pilots at this time at any rate had 4 *chio* to a *chih*, not 8, though the half-*chio* was marked on the ivory fine adjustment plate. How long before Li Hsü's time this system had been in use the text does not say. We can be sure, at any rate, that the Chinese pilots were using his method in the fifteenth century, and they may well have been doing so in the fourteenth or even the thirteenth.

Evidence indeed seems to be growing that they were taking altitudes by the beginning of the twelfth century. A text of 1124 already suggests this, and strange confirmation comes from a passage noticed by Lo Jung-Pang in the *Sung Hui Yao Kao* (Drafts for the History of the Administrative Statutes of the Sung Dynasty). Ships chartered for the defence of the Yangtze River and the sea in 1129 were to be equipped with a 'Dipper-Observer' (*wang tou*). Though we have not elsewhere encountered the expression *wang tou*, its obvious meaning is a sighting-tube for determining the positions and altitudes of the stars of the Great Bear, but the 'Dipper-observer' might equally well have been a cross-staff or *kamal*. Perhaps therefore the quantisation of stellar altitudes followed closely upon the quantisation of azimuth directions by the Chinese pilots.

Summing up the present state of our knowledge about the development of quantitative navigation in the Eastern seas, we have to start with the introduction of the mariner's compass on Chinese ships some time before 1050 A.D., possibly as early as 850. How soon this spread to the Indian Ocean we still do not know. Before 1300 there is hardly any evidence for the taking of star altitudes at sea by instrument, whether among Arabic or Indian pilots, and only very little for the Chinese navigators. But the *Shun Fêng Hsiang Sung* tells us that from 1403 'the drawings of the guiding stars were compared and corrected', which suggests a considerable previous development during the fourteenth century. Broadly speaking, therefore, we may not be far off the truth if we say that when Ibn Majid met Vasco da Gama at Malindi, fully quantitative navigation was some two or three centuries old 'East of Suez' but hardly one century old in the West.

Before taking leave of the Chinese pilots it may be of interest to glance at the contents of two or three typical rutters or navigational compendia. The first of these is the *Shun Fêng Hsiang Sung* (Fair Winds for Escort), composed by an anonymous mariner some time about 1430, at the close of the period of Chêng Ho's expeditions. The second is the *Tung Hsi Yang Khao*, compiled by Chang Hsieh in 1618, but showing no evidence of any occidental influences. The writer of these 'Studies on the Oceans, East and West' was much more scholarly as a historian and geographer than the anonymous fifteenth-century sea-captain, but seems also to have had personal acquaintance with the sea. His ninth chapter is entitled *Chou Shih Khao*, i.e. on the Ship's Master and what he should know.

Part of the introduction of the *Shun Fêng Hsiang Sung* reads as follows:

'In bygone days, the Duke of Chou discovered and worked out the principles of the south-pointing needle. Throughout the centuries from ancient times until today, these principles have circulated far and wide. Yet if you ignore the increase or decrease in the number of watches, or their divisions, you will be at fault. Thus it was that charts were drawn, and all details of voyages recorded.

'Now these old documents get worse worn every year, and it is difficult to judge from them what is the truth of the matter. If later people make copies from these originals, they will, I fear, fall into error. (So) availing myself of leisure, I have made a comparison of the calculated (number of) watches for every day, and have investigated the respective (number of) days for (each) through voyage. And I have collected and written down the number of the watches, the directions of the compass-needle, the appearance of mountains and the conditions of the water, whether there are bays and islands, shoals and deeps; with regard to all the places from the directly governed (district) of the Southern Capital (Nanking) to Thai-Tshang and Wu-Li-Yang (the Gulf of Siam, the Sumatran Seas and the Indian Ocean) (where are) the barbarian countries and other such places—in order to hand down to later generations the way and manner of making good voyages. '

Looking first at what the two texts have in common, they are found to give abundant information on landmarks and general sailing directions (*Yang Chen Lu*) with compass-bearings and soundings in fathoms (*tho*). Chang Hsieh's compendium includes as destinations Indo-China, Malaya, Siam, Java and Sumatra, Borneo, Timor, the Moluccas and the Philippines. The anonymous text goes even further afield—to Aden, Ormuz, Ceylon and Japan. Both give tables of monthly and seasonal winds (*Chu Yüeh Fêng*) with copious advice on weather-signs (*Chan Yen*) observing the shapes of clouds, the behaviour of wind and rain, together with other meteorological phenomena such as solar haloes. Both give a kind of tide table (*Chhao Hsi*), adding other signs such as the colour of the water, and any objects likely to be floating on it. Both supply the master with liturgical instructions (*Chi Ssu*), but the emphasis differs somewhat since the anonymous text is more concerned with the patron saints and tutelary deities of the compass, while Chang Hsieh likes to speak of Thien-Fei, the Mistress of the Heavens, a sailor's goddess who had, as we know, the devotion of Chêng Ho and his fleets.

The material which is only in the fifteenth-century text has some features of special interest. We hear of the method of selection of water for the floating compass, and of the proper way to make the needle float on it. Besides three tables of the 24 azimuth points, there is one listing only 14 of them as the 'palaces of the heavens', and another associating them with winds. Most interesting with regard to possible Arab influence is a small table entitled 'Principles of Star Observations', which gives the azimuth rising and setting points of four constellations, the Great Bear, *Hua kai* (significantly), the Southern Cross (*Têng lung ku*), and *Shui phing hsing* (possibly Canopus). Now such points of rising and setting were the elements of which the Arab sailor's azimuth circle graduation (lacking of course the abstract Chinese cyclical characters) was wholly constructed. Another table lists the azimuth rising points of Sun and Moon throughout successive months, with corresponding clepsydra divisions and lengths of the night and day. The fact that mnemonic verses on this, and on the Moon's times of rising and setting, follow, suggests that the pilots were widely accustomed to pay attention to these data. The anonymous text also gives mnemonic verses about lightning as a weather-sign. Lastly it adds something about the determination of currents and tides, and the calculation of watches. It is here that we find mention of the floating wood method of logging the speed of a ship.

Another anonymous rutter entitled *Chih Nan Chêng Fa* (General Compass-Bearing Sailing Directions) exists in manuscript form appended to a military encyclopaedia, the *Ping Chhien* (Key of Martial Art), by Lu Phan and Lu Chhêng-En, the preface of which is dated 1669. Besides much weather-lore and rhumb bearings for various voyages, not yet analysed, it has a sub-section on star sights (*Kuan Hsing Fa*) with diagrams of constellations. On one page, *Hua kai,* Altair (the Herdboy) and Vega (the Weaving girl), the Southern Cross and Canopus (or possibly

Achernar) can all be seen, with rising and setting azimuth points, though not altitudes, recorded at the foot of some of the columns. As we saw above, Chinese pilots were accustomed to take altitude sights of many stars apart from the circumpolars, and the stars of the equatorial lunar mansions were doubtless among them. Indeed in Chinese astronomy, the correlation of circumpolar stars with equatorial mark-point constellations was particularly prominent. Observations of the *hsiu* would give not only the time of night but also the latitude by simple calculations. It is probable therefore that further researches in the literature will bring to light nautical tables correlating *hsiu* culminations with the positions of unseen circumpolars, similar to the Arabic lists of the *manazil* described and analysed by de Saussure.

Lastly, a word on tide-tables. Since several of the extant Chinese rutters include forms of these, it is worth recalling that the phenomena of sea-tides were carefully studied in China earlier than in Europe. Authoritative histories still inform us that the oldest tide-table for a particular port is the early thirteenth-century 'fflod at london brigge', but Yen Su's *Hai Chhao Thu Lun* (1026 A.D.) contained a detailed tide-table for Ningpo. Not much later, in 1056, Lu Chhang-Ming drew up a tide-table for Hangchow which was inscribed on the walls of a pavilion on the banks of the Chhien-thang River. The Chinese pilots from the Yuan to the Chhing had thus a great tradition behind them.

The spirit of these navigators, bent primarily, even under the orders of so great an admiral as Chêng Ho, on peaceful intercourse and trade with the other inhabitants of Asia and Africa, is well seen in the concluding words of Chang Hsieh's chapter on navigation.

'According to the writer's opinion [he says], those who make carriages build them in workshops, but when they come forth to the open road, they are already adjusted to the ruts. So it is with good sea-captains. The wings of cicadas make no distinctions between one place and another, while even the small scale of a beetle will measure the vast empty spaces. If you treat the barbarian kings like harmless seagulls (i.e. without any evil intentions), then the trough-princes and crest-sirens will let you pass everywhere riding on the wings (of the wind). Verily the Atlas-tortoise with mountain-islands for its hat is no different from (an ant) carrying a grain of corn. Coming into contact with barbarian peoples you have nothing more to fear than touching the left horn of a snail. The only things one should really be anxious about are the means of mastery of the waves of the seas—and, worst of all dangers, the minds of those avid for profit and greedy of gain.'

NOTE

The author wishes to express his great indebtedness to many friends, but notably Mr. J. V. Mills and Commander David Waters, for consultation, advice and help over many years.

Select Bibliography

ORIGINAL DOCUMENTS

Azurara, G. E. de, *Chronicle of Guinea* (Hakluyt Soc., Vols. 95 and 100, 1st Series).
Bacon, Roger, *Opera Inedita* (Ed. J. S. Brewer, 1859).
Bacon, Roger, *Opus Majus* (Ed. S. Jebb, 1733).
Cosmas Indicopleustes, *Topographia Christiana* (Hakluyt Soc., Vol. 98, 1st Series).
Dicuil, *De Mensura Orbis* (Ed. G. Parthey, 1870).
Faleiro, Francisco, *Arte del Marear*, 1535.
Fernandez, Valentim, *Reportorio dos Tempos*, 1563.
Garcie, Pierre, *Le Grant Routier*, 1521.
Geographi Graeci Minores (Ed. C. Müller, 1855).
Hakluyt, Richard, *The Principal Navigations, Voyages etc.*, 1600.
Herodotus, *The History* (Trans. G. Rawlinson, 1920).
Herodotus, *The Histories* (Trans. A. de Selincourt, 1954).
Homer, *The Odyssey* (Trans. E. V. Rieu, 1946).
Homer, *The Story of Odysseus* (Trans. W. H. D. Rouse, 1937).
Kendall, Abraham, *The Voyage of Sir Robert Dudley 1594* (Hakluyt Soc., Vol. 3, 2nd Series).
Konungs Skuggfá (Trans. L. M. Larson, 1917).
Landnamabók (Trans. T. Ellwood, 1908).
Lucan, *Pharsalia*, VIII, 167–181.
Lull, Ramon, *Ars Magna Generalis et Ultima*, Lyons, 1517.
Neckam, Alexander, *De Naturis Rerum* (Ed. T. Wright, 1863).
Neckam, Alexander, *De Utensilibus* (in *A Volume of Vocabularies*, Ed. T. Wright, 1857).
Nuñez, Pedro, *Opera*, 1566.
Nuñez, Pedro, *Tratado em defensar da carta de marear*, 1537.
Peregrinus, Petrus, *Epistola de Magnete* (Ed. A. Gasser, 1558).
Periplus of the Erythraean Sea (Trans. W. H. Schoff, 1912).
Pliny, *Natural History* (Trans. H. Rackham, 1938).
Polo, Marco, *Travels* (Ed. J. Masefield, 1907).
Prüfung eines Berefferten Azimutal Compasses . . ., 1879.
Ramusio, G. B., *Viaggi*, 1554, 1556, 1576.
Regimento do Estralabio e do Quadrante, 1509(?).
Regimento do Estralabio, 1518 (?).

[Rutter in English] (Lansdown MSS. 285, No. 48).
St. Brendan, *La Légende Latine* (Ed. A. Jubinal, 1836).
Strabo, *Geography* (Trans. H. C. Hamilton, 1854).
Strabo, *Geography* (Trans. H. L. Jones, 1917).
Toleta de Marteloio (Egerton MSS. 73), 1428.
Valerius, Julius, *Res Gestae Alexandri Macedonis.*
Vincent of Beauvais, *Speculum Naturale, c.* 1296.
(*Voyage with the Earl of Cumberland*) (Sloane MSS. 3289), 1597.
Waghenaer, L., *Mariner's Mirror* (Trans. A. Ashley), 1588.
War of Saint-Sardos, 1323–25 (Ed. P. Chaplais, 1954).

SECONDARY WORKS

Barbosa, A., *Historia da Ciencia Nautica Portuguesa da Epoco dos Disco-brimentos*, 1948.
Bouguer, *Nouveau Traité de Navigation*, 1753.
Cary, M., and Warmington, E. H., *The Ancient Explorers*, 1929.
Cohen, M. R., and Drabkin, I. E., *Source Book in Greek Science*, 1948.
Cortesão, A., *Cartografía e Cartografos Portugueses*, 1935.
Diez de Gomez, G., *El Vitorial* (Trans. J. Evans), 1928.
Ferrand, G., *Introduction à l'Astronomie Nautique Arabe*, 1928.
Fontoura da Costa, A., *A Marinharia dos Descobrimentos*, 1933.
Fournier, Abbé, *Hydrographie*, 1643.
Franco, S. G., *Historia del Arte y Ciencia de Navegar*, 1947.
Gathorne-Hardy, R., *Norse Discoverers of America*, 1921.
Gatty, Harold, *The Raft Book*, 1943.
Gernez, D., Les Indications relatives aux Marées dans les Anciens Livres de Mer: *Archives Internationales d'Histoire des Sciences, 7, 671,* 1949.
Gould, R. T., *John Harrison and his Timekeepers*, 1935.
Gunter, Edmund, *Works*, 1674.
Hewson, J. B., *A History of the Practice of Navigation*, 1951.
Hourani, G. F., *Arab Seafaring*, 1951.
Jal, A., *Archéologie Navale*, 1840.
Kretchmer, K., *Die Italienischen Portolani des Mittelalters*, 1909.
Lethbridge, T. C., *Herdsmen and Hermits*, 1950.
Marguet, F., *Histoire Générale de la Navigation*, 1931.
Motzo, B. R. (Ed.), *Il Compasso da Navigare*, 1947.
Nordenskjold, A. E., *Periplus*, 1897.
Ocean Passages for the World (Hydrographic Department), 1950.
Reeves, A. M., *Finding of Wineland the Good*, 1890.
Reuter, O. S., *Germanische Himmelskunde*, 1934.
Scoresby, Wm., *Account of the Arctic Regions*, 1820.

Stoeffler, J., *Elucidatio Fabriciae ususque Astrolabii*, 1513.

Stone, Edmund, *Construction and Use of Mathematical Instruments*, 1758.

Taylor, E. G. R., The De Ventis of Matthew Paris (*Imago Mundi*, Vol. II), 1937.

Taylor, E. G. R., Hariot's Instructions, 1595 (*J. Inst. Navig.*, Vol. 5, 1952).

Taylor, E. G. R., The Doctrine of Nautical Triangles Compendious (*J. Inst. Navig.*, Vol. 6, 1953).

Taylor, E. G. R., (1) Jean Rotz and the Variation of the Compass; (2) Jean Rotz and the Marine Chart (*J. Inst. Navig.*, Vol. 7, 1954).

Taylor, E. G. R., *The Mathematical Practitioners of Tudor & Stuart England*, 1954.

Taylor, E. G. R., *Tudor Geography*, 1930.

Taylor, E. G. R., *Late Tudor and Early Stuart Geography*, 1934.

Twiss, Sir T., *The Black Book of the Admiralty*, 1874.

Index

gimbals, for suspension of instruments, 181, 183

globes, of Crates, 155; made in Lisbon, 176; spiral rhumbs on Mercator's, 177; Pedro Nuñez on, 180–1

Goa, 184

Goderick, English crusading pirate, 97

Godfrey, Thomas, his reflecting octant (sextant), 257

Golden Rule, see Rule of Three

Gomes, Diogo, on the quadrant, 159

Gonçalves, João, 183

Goodwin Sands, 133

Goths, conquer Italy, 91

Gozo, 109

Graham, George, inventor of deadbeat escapement, 260

Grand Banks, 25, 188

'Grasholm', 132

Gravesend, 200, 201

Great Belt, 102

great circle, degree of, Gunter on use in computation, 230; see also degree

great circle route, Pedro Nuñez on, 177; Dee's method of finding, 197; sailing, 233

Greece, survey of coast by Darius, 50

Greeks, and astronomy, 43–4; and navigation, 43–6; colonies, 50; trade settlements on south Russian coast, 51; use of maps, 55; familiarity with monsoon, 58

'Green Bank' (Thames), 133

Greenland, 25, 26, 76, 77, 80, 155; possible voyage by Cormac, 69; colonization of, 16, 26, 78, 79; sighted by Gunbjorn, 74; winds on coast, 75; Irish voyages to, 75; landmarks visible simultaneously with Icelandic, 79

Gresham College, 227–8; Professors, 227–8; see also Briggs, Gunter, Gellibrand

'grounds' ('bottoms') in earliest pilotbook, 132, 135, 136; in Herodotus, 35; in voyage of Pedro Niño, 142

Guadalquivir, River, 62

Guadiana, River, 62

guano, 30–1; supply dependent on absence of rain, 31

Guardafui, Cape, 61

Guiana, 219

Guinea, 174, 185, 189, 197; exploration of coast, 162

Guinea, Gulf of, 156, 162, 179, 239; and monsoon, 18, 24

Guiscard, Robert, 92

gulfs, assisting or endangering sailors, 32

Gulf Stream, 24–5, 26, 30, 246

Gulf Stream drift, 30; effects in N.-W. Europe, 26

Gunbjorn, voyage of, 74

Gunbjorn's Skerries, 74

Gunter, Edward, 228–30; death in 1626, 229; on length of degree, 230; on decimal system, 230; see also instruments

Guyot of Provins, on magnetic needle, 95–6

Guzerat, 125

Hadley, John, 257, 258, 263; his octant (sextant), 256–7

Haifa (Cayfas), 108

Hakluyt, Richard, 226; Voyages of, 196, 224

Halgoland, 84

Hall, Christopher, 207, 208, 216

Hall, Richard, 194

Halley, Edmund, 238–42; survey of wind-systems, 238–9; wind-map, 239; on charting tidal streams in Channel, 239; commissioned in Royal Navy, 239–40; isogonic chart, 240; chart of Channel, 240; appointed Astronomer Royal, 251

Hamilcar, and Breton skin-boats, 65

handbooks, merchants', 57, 81

Handson, Raphe, 225–6

Hanish Islands, 58

Hanno, voyage to Sierra Leone, 46

harbours, described by Homer, 41–2

Hariot, Thomas, 209, 215, 223; table of amplitudes, 184; on faults in Regiment of North Star, 218–19; correction tables, 219; Arcticon, 219; observation of Pole Star, 219; on instrumental errors, 220; design for back-staff, 220; on variation problem, 220–1; Doctrine of Nauticall Triangles, 223; Canon, 223; calculation of meridional parts, 223–4; map projection, 223–4; retirement from

Out-South, in Norse division of horizon, 8

Ovaéforjokul, Mount, as landmark for Iceland, 75

Oxford University, 215, 249, 260

Pacheco, Duarte, treatise on cosmography and navigation, 192

Pachynus (C. Passaro), 56

Pacific Ocean, 239; fog in, 26; currents, 26

Palestine, 107, 114, 129, 143

Paphos, 53

Papos, 72

Papyle, 72

'Paradise of Saints', island of, 67

parallax, horizontal, of the Sun, 219; stellar, 220; of the eye, see instruments (cross-staff)

Paris, 121, 186, 194, 195, 243, 257

Parthia, 56

partition of New World between Spain and Portugal, 152–3, 177

Passaro, Cape, 109; see also Pachynus

Patrick, St., and conversion of Ireland, 67

Paul, St., shipwrecked, 3, 35–6, 53

Pavia, 114

'Peccato' rock, 105

Pechora, 196

Pedro, King, of Aragon, 104

Peking, 123

peleio (long-distance crossing), 108; in Compasso da Navigare, 135

Pembroke, coast of, 132

Penmarch Point, 132, 135, 137, 138

Penryn, 140

Pepys, Samuel, 249; on need for map of ocean currents, 239; institution of naval lieutenant's examination, 251

'Peraldo', mountain above Genoa, 107

Periplus of the Erythraean Sea, 57–60, 61; of Scylax, 49, 50–1

Persian Gulf, 60

Peru, currents off coast, 30–31

Pessagno, Manuel, 157

Peter Peregrinus, on magnetic needle and Pole Star, 101; compasses devised by, 101; attempts at lodestone timepiece, 245

Phaeacia, 41

Phaeacians, as navigators, 42–3

Pharos, island and lighthouse, 63

Pharsalus, battle of, 46

Philippine Islands, 239

Phoebus Apollo, 31

Phoenicia, 40

Phoenicians, as seafarers, 43; use of Little Bear to distinguish North, 43; as colonizers and traders, 46, 50; use of star altitudes, 48; ships hired by Darius, 50; in Periplus of Scylax, 51

Phorcys, 41, 42

Phycus (Ras-al-Razat), 56

Picard, Jean, his measurement of the Earth, 231; improves micrometer, 237; his measure of degree, 237–8

'Pico', 247

Pillars of Hercules, 51, 62, 155, 156

pilotage, first mention of local, 60

pilot-books, of 3rd cent. A.D., 52; origins of 41–2; use in 4th cent. B.C., 49; first English, 131–6; contrasted with Compasso da Navigare, 131; neglect of distance in first English, 140; distance in Garcie's Grant Routier, 170; hand-copied, 168; English translation of Grant Routier, 168; Garcie's Grant Routier, 168–171; of de Castro, 184; shore profiles in, 169, 192; harbour plans in, 192; in Michael of Barnstaple (1533), 193; of Alexander Lyndsay, 194; early necessity for local, 61–2; used by Chinese, 267–8, 274–7; see also Navigational Notes and Cautions

pilot-hydrographers, officially appointed in France, 235

pilots, of Sidon, 43, 48; responsibility in ancient maritime law, 62; Arab, 125–6; Portuguese, 134; training of Portuguese, 157; of Dieppe, 185; instruction of, in Spain, 174; French, in English service, 194; instruction of, in France, 235; duties of, in Fournier's Hydrographie, 234

Pimental, Manuel, his Arte de Navegar, 231

'Pinteado, the noble', 195